Today's World
Defence

Charles Freeman

Batsford Academic and Educational Ltd
London

CONTENTS

War in the Twentieth Century	3
War and Morality	11
What are we Defending?	20
Threat Assessment	24
A Soviet Threat?	30
The British Response	42
What's Wrong with Defence Policy?	52
A Way Ahead?	61
Glossary	67
Resources List	69
Index	71

© Charles Freeman 1983
First published 1983

All rights reserved. No part of this publication may be reproduced, in any form or by any means, without permission from the Publisher

Typeset by Tek-Art Ltd, London SE20
and printed in Great Britain by
R.J. Acford
Chichester, Sussex
for the publishers
Batsford Academic and Educational Ltd
an imprint of B.T. Batsford Ltd
4 Fitzhardinge Street
London W1H 0AH

ISBN 0 7134 0969 X

ACKNOWLEDGMENTS

The Author and Publishers thank the following for their kind permission to use copyright illustrations: BBC Hulton Picture Library, for the pictures on pages 6 (top), 12, 24, 27, 30, 34, 35; Central Press Photos, pages 28 (top), 43, 46; Henry Grant, page 50-51; Imperial War Museum, pages 4-5, 6 (bottom), 14, 33; Keystone Press Agency Ltd, pages 17, 37, 45, 48, 49, 56, 65; Ministry of Defence, Army Public Relations, Berlin (photo by Karl Tietz), page 23 (top); NATO, page 22; Novosti Press Agency, page 38-39; John Topham Picture Library, pages 3, 7 (bottom), 8, 9, 18 (top), 21, 25, 26, 28 (bottom), 31, 32, 53, 58, 64; United Nations, pages 18 (bottom) (WHO photo by P.K.J. Menon), 23 (bottom), 62, 63. Thanks are also due to Pat Hodgson for the picture research on the book.

WAR IN THE TWENTIETH CENTURY

When a nation decides to go to war, it often does so at a moment of great national emotion, when what seems a vital principle of national honour or security is at stake. At such times, the use of war not only seems to be right, but often there is a strong feeling that it will be a quick and effective, even a glorious way of solving political problems.

This was the case in the First World War. When the nations of Europe declared war on each other in 1914, each believed that war would provide a solution to the disputes which had been festering among them over the previous thirty years. Within four years they knew that the reality of war was very different. A whole generation of young Europeans had been slaughtered, many simply running unprotected over open land towards machine guns. Even victors of the war, Britain, France and Italy, were scarred and economically weakened by the experience. Two of the "Great Powers" of Europe, Austria-Hungary and Russia, had collapsed completely, the first breaking up into smaller national states, while in the second a new communist society was being built, with momentous consequences for the rest of mankind.

Naturally, at the end of such a war in which millions had died, there was earnest talk of how to prevent war in the future. Many believed that no statesman would ever risk his nation in war again, when the results, even for the victors, were

National honour at stake? British paratroops leave for the South Atlantic on a mission to regain the Falkland Islands from Argentina, 1982.

so disastrous. In a determined attempt to avoid future conflict, the League of Nations was set up in 1919, as a forum where nations could meet and, through discussion, solve their problems peacefully. There was also much talk of international disarmament, and in 1928 over sixty nations signed an agreement — the Kellogg-Briand Pact — to renounce war as an instrument of policy.

And yet by 1939 all these hopes were dashed and Europe was at war again. The fascist dictators had emerged, first Mussolini in Italy, then Hitler in Germany, exploiting the humiliations and resentments of their peoples and yet again glorifying war as a way to solve political problems.

Hitler claimed the right of the German people to move eastwards, to conquer more "living

▲ The First World War saw the slaughter of a whole generation of young Europeans. Memories of the horror of this war were to scar the minds of those who survived.

room" for themselves. Britain and France, haunted by the horrors of the First World War, and determined to avoid another such conflict, hoped he could be stopped by compromise and negotiation. It was only after many hesitations that they declared war on him in 1939, after his invasion of Poland.

Germany and Italy were not the only expansionist powers in the 1930s. In South-East Asia, Japan had launched a campaign of aggression against China, a campaign which eventually brought Japan up against the might of the United States. Following a daring attack by the Japanese on the American base in Pearl Harbour, Hawaii, in December 1941, Japan and the United States were at war. A few days later, Hitler himself declared war on the United States. He had already that year invaded the Soviet Union; now he was committed to fighting the two largest nations on earth. German military might was formidable, but it was now clear that Hitler would eventually be defeated.

Not surprisingly, the destruction involved in the defeat of Germany and Japan was enormous. This was partly because the war raged over such vast areas. The Germans had conquered most of Europe and even vast areas of the Soviet Union; the Japanese had built a large empire in South-East Asia. Appalling damage was done as the Allied forces battled their way back across the conquered territory. Civilians, in particular, suffered dreadfully and often they were chosen deliberately as the main targets of bombing raids (see page 14).

The war against Japan ended in a particularly horrifying way. In the last months of the war a new and vastly more destructive form of bomb, the atomic bomb, was developed in the United States. Two Japanese cities, Hiroshima and Nagasaki, were chosen as targets. The bombs, dropped in August 1945, killed 110,000 immediately and left many thousands more to die later from the effects of radiation. Within a few days Japan had surrendered.

Looking back, it is clear that the dropping of the atomic bombs in 1945 ended one era of warfare and introduced the next. In the years that followed, four nations joined the United States as possessors of nuclear weapons: the Soviet Union, France, Britain and China. The weapons themselves developed enormous destructive power, many times that of the bombs of Hiroshima and Nagasaki. It is now possible for a

The hope of ending war for all time. Here the League of Nations meets for the first time. Within less than twenty years its dreams of maintaining peace had been shattered.

A new era of warfare dawns. An example of the first model of atomic bomb as dropped on Hiroshima.

few world leaders to unleash almost total devastation on the world community. How these developments took place and the threats they offer to mankind will be one of the main themes of this book.

War Since 1945

Despite these developments, there remained some hope in 1945 that war really could be eliminated. A new international organization, the United Nations, was set up to keep the peace and was given far stronger powers to do so than its predecessor, the League of Nations. A Security

Hiroshima after destruction in August 1945. Modern nuclear weapons contain more than eighty times the destructive power of these early bombs.

Council was set up with fifteen member nations and this Council could call on all member states to unite in punishing any acts of military aggression. Five nations — the United States, the Soviet Union, Britain, France and China — were given permanent seats on the Council and all had to agree on any resolution of this nature before it became effective. Unfortunately, the deep tensions that arose between these powers after the war have made agreement between them rare and, as a result, the United Nations has been little more effective than the League in preventing war.

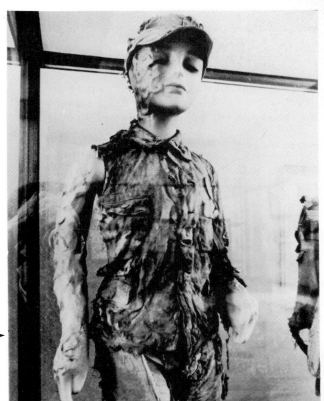

A model in the memorial museum in Hiroshima shows the effects of a nuclear attack. The figure was copied from one of the younger victims of the bomb.

Refugees from war. A refugee of the civil war in El Salvador seeks help for her child. Violence is a part of everyday life in many parts of the Third World.

It has been in the Third World and the Middle East that warfare has been most prevalent since 1945. The recent report of the Independent Commission on Disarmament and Security Issues* sums it up as follows:

Since 1945 . . . virtually all parts of the Third World have suffered the ravages of conventional warfare. For all practical purposes, war and internal conflicts have been so common and so brutal that casualty figures are almost worthless. Suffice to note that since 1945, millions have been killed directly, tens of millions have been wounded or infected with disease, and hundreds of millions have been caught in the economic and social consequences of conventional wars.

Most of these conflicts of the Third World have been internal conflicts — armed groups seeking to overthrow an unpopular government, for instance. In Africa and Asia, there have been many struggles by the native peoples of these areas to free themselves from the rule of the European nations which had colonized them. These wars have been marked by acts of brutality and terrorism by both governments and insurgents.

There have also been persistent trouble spots, of which the Middle East has been the most obvious. In 1948 the Jewish peoples of Palestine formed themselves into the state of Israel. This aroused bitter resentment from the Arab peoples of the area (the Palestinians) and the neighbouring Arab states. There have been no fewer than

*This important report is discussed in more detail on pages 62-63

A protest against the continuing threat of nuclear war. A young supporter of the Campaign for Nuclear Disarmament brings flowers to a London memorial to the victims of Hiroshima.

four major wars fought between Israel and the Arabs and numerous other outbreaks of violence. In 1982, Israel launched a major invasion of the Lebanon, adding yet another chapter of suffering to a wartorn area.

While these conflicts have continued to make war and violence an ever-present reality for many peoples of the world, North America, Europe and the Soviet Union have remained relatively peaceful. This is not to say that they have not used violence to achieve political aims. It has been estimated that the United States has used military force some 200 times and the Soviet Union some 190 times since 1945, in an attempt to achieve their political ends outside their own territories. In the case of the United States, the most prolonged and costly involvement was the Vietnam War, where, between 1960 and 1972, American forces tried to uphold a government in South Vietnam against communist insurgency. The Soviet Union invaded Hungary (1956) and Czechoslovakia (1968), when these two Eastern European nations tried to break free of Soviet control. It also launched a major invasion of Afghanistan (1979), probably to prevent this Muslim country from acting as a focus of freedom for the Soviet Union's own Muslim peoples.

Despite these conflicts, for many people the greatest threat of all to the peaceful hopes of mankind comes from nuclear weapons. Although peace between the nations of the West (the United States and Western Europe) and those of the East (the Soviet Union and China) has held, there has been continued tension between them — the so-called Cold War. Both sides in the Cold War have relied on nuclear weapons as their ultimate defence against attack. Fears that the tensions of the Cold War might erupt into actual conflict, and that this conflict might involve the use of these weapons remain very real.

These fears have led in the 1980s to an extraordinary awakening among the peoples of the Western world to the dangers of war. Hundreds of thousands of people have joined in protest against defence policies which continue to rely on nuclear weapons.

This awakening to the dangers of war has led to a vast output of literature on nuclear weapons, defence policy and disarmament. This present book cannot hope to make any startling new contribution in a field where there is already much impressive and thought-provoking writing. What it does hope is to provide a background to some of the main problems in the making of defence policy; to explain how the present situation has arisen; and to look, in particular, at how one nation, Great Britain, has faced up to the question of defence in a nuclear age. We shall start by looking in more detail at some of the more worrying developments of modern warfare.

WAR AND MORALITY

The "Just" War

No one who has studied the development of warfare in the twentieth century can fail to be aware of the far-reaching side-effects of even the most limited conflict. The course of any war is unpredictable. Violence used for whatever end cuts deep. It destroys and embitters, arouses the desire for revenge. Very seldom does any use of violence neatly achieve the objectives for which it was used.

One response to this knowledge is pacifism, the refusal to use violence in any situation, even in self-defence. Pacifism has an impressive following, especially among members of the main religious traditions, but it has always remained a minority belief. For most, there continues to be the concept of "the just war", the belief that, in certain circumstances, force, used in a limited way, can be justified.

Statesmen, church leaders and philosophers have debated the nature of the just war for centuries. In what circumstances is violence justified? How much violence is justified in achieving your objective? What rules of war should be observed to minimize the effects of conflict?

It seems at times that these discussions have got nowhere. Every nation, when declaring war, seems to be able to find a justification. It was not surprising, therefore, that, at the end of the First World War, an attempt was made to reduce the chances of all war, whether "just" or not. It was felt that the "Great War" of 1914 to 1918 had shown statesmen and their people alike that war was an irrational way of acting; the costs were always so much greater than the gains. If the causes of war could be defined, and remedies found for them, perhaps, at last, peace might come. When the outbreak of the First World War was examined for "causes", it appeared that the main triggers of war had been the system of alliances, the heavy expenditure on weapons and the hasty resort to violence at a time of tension.

So, when the League of Nations was set up, it was dedicated to the reduction of armaments, an emphasis on open discussion between nations at times of dispute, and a firm, if rather optimistic, belief that tensions would calm when the facts of a dispute were known and time had been given for tempers to cool. If aggression did take place, then the League had some sanctions, mainly economic ones, to use against the aggressor, providing that all its members could agree to enforce them.

The League depended heavily on positive leadership from the major powers. In this it was disappointed. The United States never joined. Britain and France were the leading members, but they continued to put their own national interests before those of the world community. Faced with the many conflicts of the 1930s, the rise of fascism, Japanese aggression in South-East Asia, Italy's invasion of Abyssinia and the Spanish Civil War, the League proved powerless.

When it became clear that there were leaders, such as Hitler and Mussolini, who continued to believe in and glorify war, the old question of the just war re-emerged. Accepting that all war is evil, in what circumstances is it just to wage war to remove this evil? When can an individual nation take up arms against armed aggression? For those who could remember the agonies of the First World War, it was an appalling dilemma, and does much to explain the hesitations with which so many politicians of the 1930s faced the expansion of Nazi Germany. For those on the political left, rearmament meant cutting back on welfare programmes, urgently needed at a time of world depression. Only gradually and painfully was the need to resort to violence in order to destroy fascism accepted. Julian Bell, a young Cambridge intellectual who was to die in the Spanish Civil War, wrote: "I believe that the war resistance movements of my generation will in the end

Hitler addresses a fascist rally. It was the emergence of fascism, with its glorification of war, that made many pacifists of the 1930s accept that war was justified, to destroy this greater evil.

succeed in putting down war — by force if necessary." He summed up the dilemma very well. The vast majority of the British people were united in support of declaring war on Hitler when he invaded Poland in 1939.

The debate as to when war is justified continues. The United Nations accepts a "just" use of violence in two circumstances. The first is in self-defence. The second is when a resolution has been passed by the Security Council of the United Nations calling on members to use armed force against an aggressor. When Argentina captured the Falkland Islands in 1982, Britain justified its military action in retaking them partly as self-defence (in that it was British territory that was being invaded) and partly as coming from the moral duty to punish an aggressor and deter further acts of aggression.

Justifications have also been put forward for the so-called "wars of liberation", those waged by the people of a nation against an oppressive government, on the grounds that violence is justified if no other way of gaining freedom is available. The concept of the just war is thus very much alive, although the problems of defining in what specific circumstances war is just continue to be daunting.*

*A good discussion of the issues is to be found in Michael Howard, *War and the Liberal Conscience*, OUP, 1981.

Defence or welfare? A continuing dilemma for the nations of Western Europe. Here a cartoon by the famous cartoonist Low shows Britain collapsing under the strain of trying to maintain expenditure on both.

"ALL I ASK IS THAT YOU GET IT PROPERLY BALANCED"

Moral Issues in Modern Warfare

Three specific moral issues have become central to twentieth-century warfare: the increasing suffering of civilians in warfare; the destructive nature of modern weapons; and the heavy costs of military programmes.

Civilians in Warfare

Civilians have always suffered in wartime; their crops have been razed, their homes destroyed and their lives threatened by undisciplined soldiers. The twentieth century has seen developments, mainly in the nature of weaponry, which make the lives of civilians all the more vulnerable.

The most ominous of these was the development of bombing from the air. The threat to civilians from bombing was already clear in the First World War and, in the optimism which followed the coming of peace, several nations, including Britain, promised in the Washington Treaties of 1923 that they would not bomb civilians in times of war. These ideals did not last. Already in the Spanish Civil War, the German Luftwaffe bombed civilians, notably in the

World War Two bomber crews being briefed. Casualties among crews were high, and there is little evidence that their bombing raids on German cities did much to end the war more quickly.

Bombs are loaded for another attack on Germany. Some 600,000 German civilians died in these raids.

Civilians protect themselves from bombs. A back garden in Britain houses a makeshift shelter.

Dresden, one of the great cultural centres of Germany, is reduced to ruins in the closing months of the war. Many of those killed were refugees fleeing from battles further east.
▼

notorious raid on the Basque town of Guernica in 1937. In the Second World War the policy was extended. There was heavy bombing of Britain by Germany and even heavier bombing of Germany by Britain. The United States launched similar raids on Japan. Civilians were chosen deliberately as targets. In Britain it appears that this decision was made not because it was felt that bombing civilians would be a particularly effective way of winning the war, but because the large fleets of British bombers were unable to bomb military targets with any accuracy and it seemed better to direct them against the German cities than to leave them idle. Once the new policy was accepted, it was carried on with increasing momentum throughout the war, resulting in the deaths of some 600,000 Germans.

When the first atomic bombs were tested in 1945 it was thus already accepted that cities were legitimate targets of warfare. They have never ceased to be. Most nuclear weapons are aimed directly at cities and it is civilians who thus remain the main targets of any nuclear conflict.

Civilians have also suffered in many of the conflicts of the Third World. In the so-called "wars of liberation" in which armed groups claim to be leading the people against an unpopular government, civilian casualties have been especially high. The insurgents often shelter among the civilian population, and civilians then find themselves in the centre of battle when the insurgents are attacked by government forces. Insurgent groups themselves often use violence to intimidate government supporters, with a spiral of atrocity and counter-atrocity developing as a result.

Examining these many conflicts, the Independent Commission on Disarmament Issues reports:

> Life for the inhabitants of these afflicted regions often becomes unbearable. With their villages bombed out and devastated by government troops looking for rebel forces; their food, their possessions, and their means of livelihood appropriated by rebel groups, they have little choice but to flee, preferring the unknown perils of life in makeshift camps in foreign lands to the known horrors of conventional war in their own countries. The number of refugees from military conflicts is reaching staggering proportions. According to data gathered by the UN Commissioner for Refugees, nearly eight million people now live in "temporary" camps in Africa, Asia and Latin America.

The International Red Cross has worked hard to lay down rules for the protection of civilians in warfare. Basic rules have been laid down in the Geneva Conventions of 1949, and additions to these were made in 1977, when proposals were issued for the better protection of civilians during the type of internal armed conflict described above. The recent Israeli invasion of Lebanon with its high toll of civilian casualties shows, however, how little these rules are observed. Civilians remain horribly vulnerable to the perils of modern war.

The Destructive Power of Modern Weapons
As has already been mentioned, the convention that civilians were a "fair" target of warfare was established before the dropping of the first atomic bomb. Nuclear weapons have continued to be targeted at cities, but their numbers and destructive power have increased enormously since 1945.

The bomb dropped on Hiroshima in 1945 — a uranium, 235 bomb — exploded with an energy equivalent to 12,500 tons of TNT explosive (12.5 kilotons). It killed some 70,000 people. A middle-range hydrogen bomb of the type owned nowadays in large numbers by the United States and the Soviet Union has an explosive power of some thousand kilotons — eighty times that of the Hiroshima bomb. The effects of such an explosion would be devastating. As Nigel Calder puts it in his book *Nuclear Nightmares* (Penguin, 1981):

> For a single one-megaton bomb bursting on the ground, the region of burn-out, or near total destruction, extends at least 2.6 miles in all directions. Within this region the blast, wind and innumerable fires started by the heat of the explosion smash and burn virtually all civilian structures. Closer in, within half a mile of the explosion, the destruction is beyond comprehension: the blast pounds the strongest

buildings like a giant hammer and bursts people's lungs, the radiant heat consumes flesh and the nuclear radiation is a thousand times the fatal dose. Thus citizens in the inner region are in effect killed in three ways at once. The crater, where everything is vaporized, is about the size of a football stadium. Particles of soil laced with radioactivity are blown by the wind to settle as dangerous fall-out over hundreds of square miles. If the attacker chooses to explode his one megaton bomb high in the air as an 'airburst', there is no crater and no local fall-out but the area of burnout is nearly three times greater — 60 square miles. That is for an unremarkable nuclear weapon — standard issue, one might say, and weighing only half a ton or thereabouts.

This would be the effect of one bomb. At present, the United States has some 9,800 strategic (or intercontinental) nuclear missiles and some 20,000 tactical (or short-range) nuclear weapons. The Soviet Union has about 7,000 strategic weapons and 15,000 tactical weapons — fewer than the United States but, in the case of the strategic weapons, on average much more powerful. In total, it has been calculated that there are now the equivalent of one million Hiroshima bombs in existence, or three tons of TNT for every man, woman and child on the earth.

The United States and the Soviet Union account for some 95 per cent of the world's nuclear arsenals. Three other nations, Britain, France and China, own the remaining 5 per cent. There are fears that other nations will acquire nuclear weapons. In 1974 India exploded a nuclear device, and it is widely believed that Israel has advanced nuclear capabilities. There are probably some eight other "threshold" nations, those with experience of working with nuclear materials.

In an attempt to stop the spread of nuclear weapons, a number of nations, including the United States, the Soviet Union and Great Britain, signed a Non-Proliferation Treaty in 1970. The intention of this treaty was to restrain any new nations from joining the nuclear club. Non-nuclear nations were to be offered nuclear technology for civilian purposes (in energy programmes, for instance) and, in return, these nations promised not to develop bombs. The nuclear nations said they would take steps to reduce their own stocks. In fact, they have not done so. Two of the nuclear nations, France and China, did not even sign the treaty. Thus the treaty has major flaws and does not seem to offer a really effective restraint on further proliferation.

Nonetheless, some comfort can be derived from the fact that there has been no nuclear explosion since that of India in 1974. The danger of proliferation seems less than it did twenty years ago. Even so, the dangers of the introduction of nuclear weaponry into such areas of instability as the Middle East are obvious.

A Trident missile — one of the new generation of nuclear weapons. It carries multiple warheads, delivered, it is claimed, with great accuracy. But is there any point in yet more nuclear weapons?

The Cost of Modern Weapons

Money spent on weapons and other military programmes is money wasted. The weapons either lie idle or are used in destruction. The money spent

A Royal Navy Sea Harrier as used by British forces in the Falklands crisis, 1982. The cost of such planes is enormous, especially when placed against the needs of the Third World.

For the price of one such plane, thousands of dispensaries like this could be set up.

on them is wasted at a time when the needs of humanity have never been so obvious or so urgent. It is estimated that world spending on arms in 1982 will amount to some 650,000 million US dollars. This is more than the entire combined income of 1,500 million people living in the fifty poorest countries of the world.

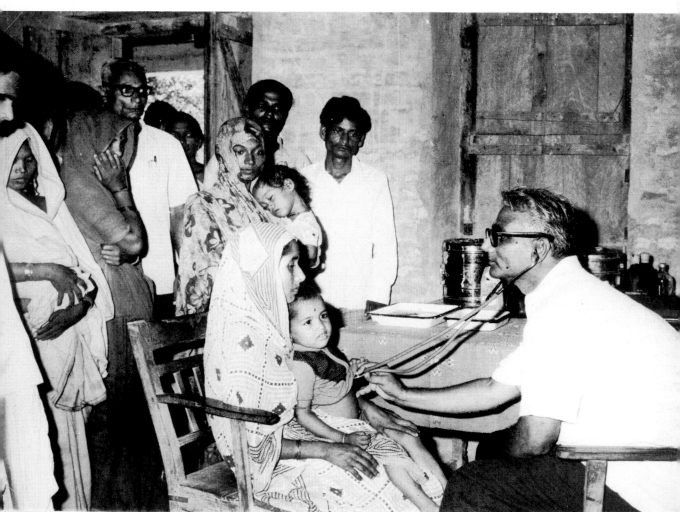

The Brandt Report on International Development, issued in 1980, showed dramatically how the most urgent needs of mankind were being forgotten as ever greater sums were spent on arms.

1. The military expenditure of only half a day would suffice to finance the whole malaria eradication programme of the World Health Organization, and less would be needed to conquer river blindness, which is still the scourge of millions.

2. A modern tank costs about one million dollars; that amount could improve storage facilities for 100,000 tons of rice and thus save 4000 tons or more annually; one person can live on just over a pound of rice a day. The same sum of money could provide 1000 classrooms for 30,000 children.

3. For the price of one jet fighter (20 million dollars) one could set up about 40,000 village pharmacies.

4. One half of one per cent of one year's military expenditure would pay for all the farm equipment needed to increase food production and approach self-sufficiency in food-deficit low-income countries by 1990.

There is naturally enormous variation in the amount of money spent by different states on their defence policies. While some Middle East nations spend over 20 per cent of their national income on defence, India spends less than 3 per cent and most Latin American countries less than 2 per cent. This compares with 5 per cent spent by Britain and the United States and 11 to 15 per cent estimated for the Soviet Union.

Much of the weaponry for Third World nations is provided by the military superpowers. According to the Stockholm International Peace Institute (SIPRI)*, the Soviet Union is the world's largest supplier of arms, followed by the United States, France and Britain. The volume of these sales appears to be increasing dramatically. While world trade in general in the 1970s rose by 70 per cent, SIPRI reports that the volume of sales of weapons to major Third World countries went up by over 300 per cent.

This policy of arms sales seems extraordinarily shortsighted. In many cases, the suppliers of arms have been caught out when what seemed to be an ally, worthy of help with its defence, turned out, after a change of government, to be an enemy. For Britain, the full absurdity of the situation became clear in the Falkland Islands dispute with Argentina in 1982: the British task force expedition to the Falklands faced weapons supplied to Argentina by Britain's allies, Germany, France and the United States, as well as by Britain itself.

Conclusion

The concept of the "just" war has always rested on the ideal that, if war is used, its effects must be limited as far as possible. The developments of modern technology have shown just how vulnerable this ideal has become in the modern world.

In recent months, much of the discussion on the "morality" of modern warfare has centred on the possession of nuclear weaponry. It has been argued that any reliance on nuclear weapons, even if they are held only as a threat, is inherently immoral. A nation possessing such weapons must thus abandon them irrespective of whether its enemies do so.

Opponents of this view would argue as follows. The fundamental "moral" duty of a nation is to protect its citizens. In a nuclear age, the most "moral" policy is that which offers least risk of nuclear attack. If (and, of course, only if) you believe that the possession of nuclear weapons to be used in retaliation if attacked will significantly reduce the risk of being attacked in the first place, then it can be seen as moral to possess them. This argument would not, of course, explain why retaliatory nuclear weapons need to be aimed at civilian targets nor why more than a very few such weapons need to be possessed, when the destructive power of each is so great.

*SIPRI, 13th Annual World Armaments and Disarmament Yearbook, 1982.

WHAT ARE WE DEFENDING?

Most of this book will be concerned with developing the themes discussed so far. But, first, there are two central problems of defence policy which need to be considered. The first can be summed up in the question: "What are we defending?" Assuming that a nation is prepared to maintain armed forces to protect itself, in what circumstances should it be prepared to use these forces?

The traditional rules of international law and the Charter of the United Nations both accept that a nation has the right to defend its own territories against attack from outside. It thus has the right to maintain an army, or in the case of an island, a navy, to fight off any such attack. This right is unquestioned.

In the past forty years, however, the development of nuclear weaponry has reduced the effectiveness of traditional forces — armies and navies — as a means of defence. A nuclear attack could, within a few moments, wipe out the major cities of a nation; and no form of effective defence against nuclear attack has yet been devised. For nations such as Britain, which have sheltered for centuries behind the defences of the sea, this sudden vulnerability has led to a complete rethinking of defence policy.

Assuming that nuclear weapons will continue to exist and that no effective means of defence against them can be devised, the central question of defence policy becomes how best to avoid attack.

At present, Britain and its Western allies have evolved the policy known as *deterrence*. The concept of deterrence is based on the belief that, if you make the risks of an attack on your territory far higher than any possible gains which an attacker may make, then he will be discouraged from attacking. In the present strategies of deterrence, the "risks" are provided by nuclear weapons aimed at the potential aggressor's territory. In the Cold War, for instance, it has been made clear to the Soviet Union that if it attacked what was thought to be a vital Western interest, it would run the risk of being attacked by nuclear weapons. The hope is that this will prevent the Soviet Union considering any attack. The concept of deterrence has grave flaws (see page 56) and certainly cannot offer any long-term security for mankind. The risks of a continued reliance on the concept of deterrence are certainly high.

So too are the risks offered by any policy which has no effective armed force to back it. If a nation has no means of retaliation, it is very much at the mercy of any other nation prepared to use weapons, in particular nuclear weapons, to achieve its ends.

In short, the coming of the nuclear age has brought a new insecurity into defence policy making. Experts are faced with the problem of weighing up which of several options provides the least risk of nuclear holocaust. If they were to miscalculate, the results of failure would be no less than the destruction of the world community as we know it. It is little wonder that the debate over nuclear policy has been so vigorous and often so bitter. Without effective international control over nuclear weapons, these appalling dilemmas will continue.

The debate over the most effective means of national defence in a nuclear age is bound to continue. It is also worth mentioning that many nations maintain responsibility for the defence of territories outside their own borders. In some cases, these are overseas possessions acquired by the European powers in previous centuries and which, for one reason or another, have not become independent nations. Britain, for instance, is still responsible for the defence of the Falkland Islands in the South Atlantic, Gibraltar, at the entrance to the Mediterranean, and Hong Kong off the mainland of China. Such possessions offer particular problems of defence. They are often many thousands of miles away and expensive to defend in any realistic way. The British recapture of the Falkland Islands, after their invasion by Argentina in 1982, was a remarkable military success, but showed only too clearly how difficult effective defence of such possessions is to maintain.

The most common form of shared responsibility for defence comes in the long-term military alliances which have been a feature of world politics since 1945. The two most important of these alliances have been the Warsaw Pact, an alliance of the Soviet Union with the nations it controls in Eastern Europe, and the North Atlantic Treaty Organization (NATO), an alliance of the United States and Canada with the states of Western Europe. Both these alliances grew up in the very special circumstances of the Cold War (see page 32). They arose partly because the hopes that the United Nations would provide genuine security for mankind proved false, and so security was sought in regional alliances, each member guaranteeing to protect the security of the others. Secondly, the enormous costs of modern defence policies made it much more realistic for nations to join together to share these costs. These two particular alliances, NATO and the Warsaw Pact, are also concerned with defending particular "ways of life" — the capitalist or democratic way of life against the socialist alternative.

The North Atlantic Treaty Organization was founded in 1949, when the nations of Western Europe decided to join with the United States and Canada, to form a more coherent defence against possible further expansion of the Soviet Union into Europe. (The basis for these fears is described in more detail on pages 30-41.) An attack on one member of the alliance was to be regarded as an attack on them all, and they would cooperate in developing joint military plans. For nations such as Britain, this was a major reversal of traditional defence policy. For the first time in its history, it was to maintain a permanent army on the mainland of Europe and, in fact, the borders of East and West Germany became the frontline of its own defence.

NATO is what we call an inter-governmental body. This means that, although there is central coordination of defence plans, no member loses its independence or ultimate control over its own defence policy. Britain was thus perfectly within its rights to detach ships and troops from NATO forces to send them to the South Atlantic to recapture the Falkland Islands.

Britain still retains responsibility for the defence of a few far-flung remnants of its Empire. Here British troops land in Hong Kong — in this case, however, to help curb illegal immigration.

The member nations of NATO retain control of their own defence policy but cooperate when they can. This is a multi-national destroyer force on exercise in the Atlantic.

Commitments to long-term alliances of this kind have their advantages and disadvantages. There is, of course, the advantage of spreading defence costs among member nations, and allowing each member nation to concentrate on the role for which it is best suited. Britain, for instance, has continued to maintain a strong navy.

The disadvantages are that decision-making is necessarily more complex and may become muddled at a time of crisis. There may well be fundamental differences between members, in their assessment of the threats facing them and the nature of the response demanded. This is certainly true of NATO, which is greatly dominated by the power and influence of the United States. The United States sees itself as the leader of a crusade against the spread of Soviet communism. It has tended to take this threat more seriously than its European allies and this difference in perception might prove serious in a time of crisis. In the Middle East crisis of 1973, the United States put all American troops in Europe on alert, without any consultation with its European allies. If the Soviet Union had reacted to this alert and a crisis had escalated, the European nations could have found themselves dragged into a major confrontation, due to the independent action of their largest member.

Sailors from six different member states of NATO. They were all working on the multi-national destroyer force shown in the picture above.

Since 1980, and the election of Ronald Reagan as President of the United States, there has been increasing concern among America's allies as to the direction of its defence policies. There have been large increases in American defence budgets. In addition, statements from the President and his advisers on nuclear weapons have suggested confusion and, to some, even recklessness, over their possible use at a time of crises. One response has been a renewed interest in the possibility of Europe "decoupling" from the United States and developing an independent defence policy.

One major drawback to this would be the expense. At present the United States provides two-thirds of the total costs of NATO and its withdrawal would lead to a heavier defence burden for the remaining members. They would certainly not find it easy to carry these extra costs.

To sum up, any nation planning a defence policy will have to make provision for a possible attack on itself, attacks on overseas territories for which it is responsible, or attacks on any nations to which it is allied.

In this chapter we have so far assumed that a nation has total responsibility for and control over its own defence policy. It decides what threats there are to its interests and what forces are needed to face them. The effect of this has been that, if one nation increases its defence spending, its opponents respond, in order to maintain their sense of security. While one nation may temporarily feel more secure, continued responses of this nature simply lead to a spiral of arms buying and increased insecurity for mankind as a whole.

It is clear that, in any rational world, a system of collective security would be worked out, by which world order would be effectively maintained by some form of world government able to punish aggressors, so that individual nations would not have to rely on their own efforts.

Some provision for this has been written into the Charter of the United Nations. Every member of the United Nations agrees, under Articles 24 and 25 of the United Nations Charter, that it will carry out the decisions of the Security Council on whom the members "confer primary responsibility for the maintenance of international peace and

British tanks in West Berlin — a continuing reminder that the front line of the defence of Britain has become Western Germany.

The Security Council of the United Nations in session. Here the Falkland Islands crisis of 1982 is being discussed. Despite the efforts of the UN the dispute was still solved by force.

security". Unfortunately, the members of the Security Council to whom this responsibility has been given have all too often been unable to agree. The United Nations has not, in practice, been a very effective force for peace. The need for an effective form of collective security will be discussed in more detail on pages 64-65.

The second central problem of defence policy is what we might call threat assessment. How does a nation assess the nature and danger of threats to its security? This will be the subject of the next chapter.

THREAT ASSESSMENT

At the beginning of April 1982 there appeared to be no immediate threats to Britain's national security. Yet, only a few days later, an angry nation decided it must send off a major naval task force in an attempt to recapture a group of islands several thousand miles away in the South Atlantic. Argentina had invaded the Falkland Islands, in pursuit of a long-standing claim to the territory, and this caught the British Foreign Office completely by surprise. Lord Carrington, one of the most able and experienced Foreign Ministers Britain had had for some years, was obliged to resign.

The unexpectedness of the Falklands crisis shows how difficult it is to make accurate assessments of the threats to a nation's national security. And yet the major decisions of defence policy, what percentage of national resources to spend on armed forces, what forms these forces should take, and where they should be stationed, must all depend on a correct assessment of potential threats. The position is made much more complicated by the need to plan ahead. The weapons of the 1990s must often be ordered in the early 1980s, so it is not just present threats which must be assessed, but future ones.

A study of British foreign and defence policy in the twentieth century shows how easy it is to make the wrong assessment. Hitler's coming to power in 1933 in Germany, for instance, posed a major problem for British defence experts. Hitler had made it quite clear, in his writings and speeches, that he believed in Germany's right to expand eastwards, to conquer new "living room" for the German people. Once in power, he began to rebuild the military and economic strength of Germany to achieve this end. Yet it took some time for the true nature of the threat to become apparent to the majority of British politicians.

Why was this so? At the end of the First World War, in the excitement of victory, the blame for

Neville Chamberlain returns from his meeting with Hitler at Munich in September 1938. Chamberlain was convinced that he had ended the threat of war, but Hitler broke the agreements within a few months and by September 1939 Britain was at war with Germany.

the war had been put on Germany and its allies. At the Peace of Versailles, Germany was forced to pay reparations, or damages, reduce its armed forces and surrender its colonies and some of its border territories. As feelings cooled, many, especially in Britain, felt that Germany had been too harshly treated. They believed that Hitler wanted little more than to right the injustices of the Versailles settlement. Once this had been done, they argued, Nazism would lose its aggressive nature and be tamed.

There was also a deep-rooted and morally impregnable argument that a negotiated settlement of Hitler's grievances would be infinitely better than a confrontation, which would renew the risk of war. Many people were haunted by the horrors of the First World War and memories of how easily the nations of Europe had slipped into war. Neville Chamberlain, the Prime Minister of Britain from 1937 to 1940, saw himself with a mission to bring peace to Europe, and he was prepared, with the support of many others, to work energetically to achieve a peaceful settlement of Europe's problems, through discussion and compromise.

Sadly, the "appeasers", as they were known, failed to recognize the total ruthlessness and insincerity of Hitler. They failed, too, to see that Nazism was a new political phenomenon which not only glorified war but depended on it for maintaining its momentum as a political force. The high point of appeasement came in the Munich Agreement of 1938. Here, Britain and France allowed Hitler to have the border area of Czechoslovakia, on the grounds that its population was largely German. They believed that Hitler would now be content. Once the border was handed over to Hitler, Hitler was, in fact, left free to expand eastwards into the now unprotected Czechoslovakia — which he did in the following year. It was only then that Britain and France realized how wrong they had been. They promised Poland that if it was attacked, they would declare war on Germany in its support. It was to keep this promise that they came into the Second World War in 1939.

It could be argued that the failure of this appeasement of Hitler has influenced British defence policy ever since. This became clear in the first major post-war crisis for Britain, the nationalization of the Suez Canal Company (of which Britain and France were major shareholders) by President Nasser of Egypt in 1956. The British Prime Minister at the time, Anthony Eden, had a long and distinguished record in foreign affairs. He had assessed the true nature of Hitler in the 1930s and had, in fact, resigned his government post rather than accept the policy of appeasement. This and his subsequent successful experience as

German troops march into the German-inhabited parts of Czechoslovakia, in accordance with the Munich Agreement. Hitler was now in an excellent position to move forward into the rest of Czechoslovakia, as he did the following year.

Foreign Secretary in the early 1950s gave him immense prestige within his party, the Conservatives. When Nasser nationalized the Canal Company, Eden saw the situation as a repeat of that he had known in the 1930s. Here was the new Hitler and he must be resisted at the first opportunity.

Eden made a secret agreement with the French and the Israelis by which they and Britain would invade and recapture the Canal. Militarily, the adventure had every chance of being a success; politically, it was a disaster. Even the United States, normally the most staunch of Britain's allies, refused to support the action. Not only had Eden made a major political miscalculation; he had acted against the whole spirit of international

Sir Anthony Eden, one of Britain's most distinguished Foreign Secretaries, who, when Prime Minister, made a fatal political miscalculation in his handling of the Suez Crisis of 1956.

law. Nasser was certainly within his rights to nationalize the Company, as long as he paid compensation to the shareholders and kept the Canal open to international shipping. Nor could it be argued that Nasser was another Hitler. Egypt had neither the economic nor the military strength to maintain a policy of expansion. The British and French troops eventually withdrew under international pressure and Nasser's prestige soared. Eden, an ill and broken man, resigned the following year.

This sad episode shows once again how easily the nature of a "threat" may be misread, even by a highly experienced politician. Bearing this in mind, how can we begin to assess the threats against which British defence policy in the 1980s must be planned?

It is first important to try to define some of the situations in which armed conflict is likely to break out. One of the most important forms of tension in the twentieth century has been the conflict between different political or economic systems. Such a clash was seen at its most bitter in the Civil War in Spain between 1936 and 1939, when the forces of conservatism, the Church, the army and the richer classes battled against the forces of the Republican government, which represented the working-class parties, the communists and socialists, and their supporters.

The clash between different political systems, with beliefs in very different "ideal societies", forms a major element in the Cold War confrontation. The communist states claim that they are on the way to building the perfect society, where all members will live in equality. The capitalist societies of the West lay the main emphasis on individual freedoms. They would claim that communism has destroyed individual freedom in its attempt to enforce an "equal" society which can never exist in practice. At the beginning of the Cold War, back in 1947, President Truman of the United States summed up the American view:

> Our way of life is based upon the will of the majority and is distinguished by free institutions, representative government, free elections, guarantees of individual liberty, freedom of speech and religion and freedom from political oppression.
>
> The second way of life is based upon the will of a minority forcibly imposed upon the majority. It relies on terror and oppression, a controlled press and radio, fixed elections, and the suppression of personal freedoms.
>
> I believe that it must be the policy of the United States to support free people who are resisting attempted subjugation by armed minorities or by outside pressures.

The "second way of life" was, to the President, that of the Soviet Union under communism. Truman set the pattern of American foreign policy, by which the United States was to have a mission, in the words of a later President, John Kennedy, "to spread the disease of liberty". The most recent United States President, Ronald Reagan, has used much the same words when outlining his policy towards the Soviet Union.

This is, of course, an oversimplified view of world politics. In the Vietnam War, the United

The Suez Invasion of 1956. British troops with civilians in Port Said. Militarily, the invasion, by British and French troops and in collaboration with the Israelis, was a success. Politically, it was a disaster and was condemned by world public opinion.

States found that upholding "freedom" often meant supporting a government which had little popular backing, and conducting a war in which the main casualties were civilians of the nation which was being kept "free". The oversimplification does not alter the fact that disputes between different political systems are one of the great dividing forces in world politics and a catalyst of international suspicions.

Britain, without necessarily supporting the more simplistic interpretations of world politics offered above, is, of course, firmly in the "democratic" camp. Like most nations of Western Europe, Britain can legitimately claim to have maintained a wider range of freedoms for its peoples than most societies in the world and, naturally, it is concerned to protect them. The real question, to be discussed in the next chapter, is how far Britain's way of life is, in fact, under threat from Soviet communism.

While the conflict between the different political systems of the East and the West has remained the most obvious source of tension in the international community, actual war has broken out for several other reasons. One of the most common forms of conflict has been the "wars of liberation" fought *within* states. The most common form of dispute *between* states seems to have been over territory. The Middle East, probably the most unstable area in the world, is the best example of this. Both Palestinians and Jews claim the same territory, the Jews on the basis that it was their ancient homeland, the Palestinians because they had occupied the area more recently and were displaced by Jews who had emigrated to the area and then set up the state of Israel in 1948. No fewer than four wars have been fought over the issue since then. The Falkland Islands is another such disputed territory. It is certainly true to say that disputes over territory tend to be deep-rooted and emotional. They provide some of the most intractable problems of world politics.

27

South Georgia in the South Atlantic — a barren piece of land — but, as the Falkland Islands crisis showed, even such territories as this arouse deep-rooted emotions.

Crowds in Argentina applaud the capture of the Falkland Islands by their troops in April 1982. Was the Argentinian government seeking to divert attention from the serious economic and political discontents at home?

The Likelihood of War

Once the defence experts of a state have asked what potential sources of dispute exist between their nation and others, the next step is to assess how far these tensions are likely to escalate into violence. How can we predict when a nation is actually prepared to use force to achieve its ends?

Some nations glorify war in itself as a means of solving political problems. Some political systems, such as fascism, see war as the real test of a nation's strength, a necessary ritual by which national identity is achieved. We see the same emotion expressed by some of the writers on the wars of liberation. They talk of the colonized nations achieving their national identity by the very act of violence which liberates them from those who have colonized them.

In other cases, a nation may have successfully used military force in the past and thus comes to regard it as a positive part of its foreign policy. The successful use of the Prussian armies to unify Germany, in their crushing defeats of Austria and France in 1866 and 1870, gave the military a strong voice in German politics in the years to come. After the restraining hand of Bismarck, the German Chancellor, was removed in 1890, military considerations increasingly influenced German foreign policy.

Every nation varies in the extent to which military leaders influence the course of foreign policy. In many nations, the armed forces are kept firmly subordinate to the civilian government. In some cases, on the other hand, military leaders are the government. A good example of the conflict over control between civilian and military can be seen in Japan in the 1930s. The first major act of aggression by Japan, the take-over of Manchuria, was the result of independent action by the army, which the civilian government was powerless to prevent. From that time on, the military had increasing influence in the conduct of Japanese policy and, with its glorification of war, led Japan down the dark valley to ultimate catastrophe in 1945.

In some cases, military action is used in an attempt to unite a nation at a time of internal crisis. The hope is that a successful war will calm national divisions. It has been argued that by taking over the Falkland Islands in 1982, the Argentinian government was attempting to use military victory as a diversion from economic weaknesses at home. Certainly, military action is enormously popular if successful, but very seldom does it have the neat results its users would hope for.

It is the main job of defence and foreign affairs experts to evaluate the nature of the governments with which they have outstanding disputes and to assess how far they might be tempted to solve these disputes by military means. All the factors we have mentioned above may be relevant in conditioning the response of a nation in time of crisis. This is the vital question. At a time of crisis, how readily will a nation resort to the use of its armed forces?

If there is considered to be a high risk that a nation will resort to force to achieve its ends, there still remains the question of how best to respond to this threat. One way forward is, naturally, to try to defuse any crisis by diplomatic means. Another is to try to rally international support, possibly through the United Nations, in the hope that by bringing the influence of the international community against a potential aggressor, the aggressor will be forced to step down. This is not an approach which has proved very successful. Again, it is possible to hope to deter attack by maintaining effective military force which would be used in retaliation. There are grave weaknesses in this approach, as we shall see (pages 56-57), but it remains one of the most common methods of attempting to avoid conflict.

The problems involved in assessing the nature of threats, against which a defence policy must be built, are awesome. We should not be too surprised that so many mistakes have been made in the past. Perhaps the only lesson which is at all clear is that defence policy must always remain flexible and prepared to deal with the unpredictable.

Since the Second World War, British defence policy has been set in a groove unlike any it has known before. A threat has been seen as coming from one nation, the Soviet Union, and this threat has been judged consistent and serious enough for the major part of the British defence programme to be geared to meet it. In the next chapter we shall consider this "threat" in more detail.

A SOVIET THREAT?

The Emergence of Communist Russia

Fear of Russia is not a new factor in British foreign policy. Throughout the nineteenth century, Russia was seen as a major threat to the British Empire in Asia — above all, India — and the approaches to it. It seems certain that these fears were exaggerated, partly, it is said, because the small-scale maps used by the British Foreign Office made the Russians look closer to British interests than they actually were! In fact, despite its vast size, Russia was not a strong nation in the nineteenth century and was increasingly unable to compete with the smaller but more efficient industrialized nations of Western Europe.

By the beginning of the twentieth century, British fears of Russia had diminished, and in 1907 the two countries made a treaty of friendship in which they settled their outstanding disputes. They fought as allies in the First World War.

The war was, however, the final blow to the weak Russian political system and in early 1917 it

Unrest in Russia in 1917. This picture shows one of the many disturbances in Petrograd before the Bolsheviks seized power in October of that year.

German soldiers, retreating from Russia in the Second World War, force Russian peasants to march with them. The German invasion of the Soviet Union brought appalling sufferings to the Russian peoples.

collapsed. After Russia had drifted for some months under a Provisional Government, a small but determined group of revolutionaries, the Bolsheviks, managed to seize power. The Bolsheviks' ability to retain power in the years to come was remarkable. They faced counter-attacks by remnants of the old government, moderates and other revolutionary parties they had pushed aside. These opponents were joined by the armed forces of several nations, Britain, France, the United States and Japan among them. The aims of these outsiders were confused and the expeditions they sent small and militarily insignificant, but they left deep-rooted suspicions in the minds of the Bolsheviks. In museums of history in the Soviet Union (as we normally call Russia after the revolution) today, this intervention is prominently displayed.

The Bolsheviks finally defeated their enemies and, under their leader, Lenin, consolidated power, laying the foundations for the world's first communist state. They had hoped that their revolution would spark off others in the rest of Europe, but this was not to happen. By the late 1920s and early 1930s, the Soviet Union, under the leadership of Josef Stalin, was concentrating on the building of "socialism in one country". The Soviet Union continued to remain leader of the world communist movement and insist that world revolution would inevitably come, but its main effort was concentrated on the building of an impressive industrial power base. These were years in which the Soviet Union was certainly viewed by many with suspicion, but not with any real fear.

It was in the mid-1930s, with the rising power of Hitler and his declared intention of expansion towards the East (and thus towards Russia) that Stalin began to come out of isolation. He attempted to build a "Popular Front" with France, Britain and other anti-fascist nations, against

31

Hitler. Mutual suspicions, however, doomed this to failure. Stalin had intervened in the Spanish Civil War, and it soon became clear that he was hoping to engineer the triumph of the Russian-controlled Spanish Communist Party in this war. Reports were also reaching the West that the Soviet army was being ruthlessly purged of many of its leading officers and thus it was right to ask whether it would offer much effective help if war came. Stalin, for his part, was shocked by the Munich Agreement by which Britain and France allowed Hitler the borders of Czechoslovakia, thus increasing the vulnerability of the Soviet Union. In an attempt to stave off war, Stalin signed an agreement with Hitler in 1939, by which he agreed to remain neutral when Hitler invaded Poland. In return, he was to receive some of the spoils, including Eastern Poland, for himself.

Hitler treated Stalin as ruthlessly as he did everyone else. In 1941 he launched a massive invasion of the Soviet Union, leading to four years of quite appalling suffering and devastation for the Russian people. In the same year, Hitler declared war on the United States. Britain was still unconquered, and there was now formed a strange alliance of the Soviet Union, the United States and Britain, united in their determination to overthrow Hitler but with little else in common.

Stalin, in particular, retained deep suspicions of Britain and the United States. He believed that they wished him to bear the brunt of the war, so that, when victory came, the Soviet Union would be exhausted and they would be able to reap the fruits of victory. He also feared that they might come to terms with Hitler before he was defeated, with the result that fascist states would remain in Europe. Britain and the United States recognized these fears and announced that they would fight on until the Germans (and Italians) surrendered unconditionally. They also launched two invasions of Europe, in 1943 and 1944, one through Italy and the other through France.

The Coming of the Cold War

As the war came to an end in 1945, Stalin's armies occupied most of Eastern Europe. Some of these areas were ones he had gained through his "deal" of 1939 with Hitler — Lithuania, Latvia and Estonia and Eastern Poland; others —

As the Russians cross into Poland towards the end of the Second World War, Polish peasants flee before them.

The Yalta Conference of 1945. Here we see Churchill, Roosevelt and Stalin discussing the future of Europe. Churchill and Roosevelt believed that Stalin had promised free elections in Poland. These were never to take place.

Hungary, Bulgaria and Romania — had been allies of Hitler. Britain and the United States knew they would not be able to dislodge Stalin from the areas gained in 1939, and they had little immediate sympathy for the allies of Hitler. It was his occupation of the whole of Poland which they found most unforgivable. After all, it had been to maintain the independence of Poland that Britain and France had first declared war on Hitler. For the West, therefore, the acceptance of Soviet control over Poland would be a humiliation. They clung to promises which Stalin made early in 1945, at the Yalta Conference, that he would allow free elections for the Poles, but these never took place. The only elections which did take place, two years later, were a farce. The Western allies could do little. They had neither the military means nor the political will to dislodge Stalin. Stalin, for his part, saw control of Poland as vital for his security. In both World Wars, Russia had been invaded through Poland. Poland had to be maintained as a buffer state to protect the Soviet Union if Germany was ever to regain its strength.

Disputes also arose over Germany. The defeated Germany was split into four zones, with the Russians taking the eastern one, and France, Britain and the United States the three western ones. The long-term plan was that Germany would be reunited. However, the Soviet Union consolidated its grip on the eastern zone and, gradually, what was supposed to be a temporary division of Germany became a permanent one.

There remained the question of the former German capital, Berlin. Berlin was in the eastern, Russian, zone, but it had been partitioned between the Allies, in the same way as the rest of Germany. There thus remained a West Berlin, controlled by Britain, France and the United

33

States, *within* Soviet-controlled territory. This was a great embarrassment to the Russians. It was clear that the Western Allies would be determined to maintain their foothold in Berlin and, after a successful currency reform in the western zones of Germany, which was later extended to western Berlin, this part of the city appeared as a symbol of capitalist prosperity in what remained a shabby and impoverished eastern Germany. In 1948, Stalin, in a dramatic move, closed off the access routes to Berlin, hoping to force it into surrender. The West was forced to make a reaction. There was some talk of military retaliation. Finally, the decision was made to try to keep Berlin supplied by air. For several months, thousands of tons of supplies were flown in. By the spring of 1949, Stalin knew he had been defeated and he reopened Berlin to the West.

Another extension of Russian power took place in 1948, this time in Czechoslovakia. In elections held after the war, the Communists had secured a third of the votes and held several of the ministries in the Czech government. In February 1948, a political coup was engineered and non-communist members of the government were ousted.

In 1949 news came that the Soviet Union had detonated its first atomic bomb.

These were the events which formed the background to the Cold War. Over the past thirty years historians have picked over every event of these years, trying to assess the motives of the statesmen involved and to apportion blame. It is only recently, however, that many of the government archives for the period have become available and a full assessment is becoming possible.

There is no doubt that Stalin was obsessed with his need for security. Suspicious by nature, he had little trust in the Western powers. They had betrayed him before the war and, in his eyes, had allowed Russia to bear the brunt of the fighting when war came. He could never allow the eastern borders of Russia to be vulnerable to another devastating attack as had just been endured.

Like so many statesmen before him, Stalin became a prisoner, not only of his own fears but of the past. He failed to realize how different conditions in Europe had become. Western Europe (in particular, his old enemy, Germany) was

In 1948 Stalin closed off West Berlin to the West. It was decided to keep it supplied by flying in materials. Here supplies are loaded.

shattered by the war. Stalin was offered an equal share in determining Germany's future. There was little to fear from the other European nations. Admittedly, the United States had the atomic bomb (and had shown that it was prepared to use it), but, traditionally, the United States had followed a policy of isolation, and it was clearly in no mood to challenge the Russians. American military spending plummeted in these years from 20 per cent of its national income in 1946 to 4.6 per cent in 1950.

Stalin's aggressive takeover of Eastern Europe, with his consolidation of power in Czechoslovakia and his attempt to winkle out the British, French and Americans from West Berlin, had the effect of achieving a major reversal in American foreign policy. From being an isolationist power, the United States gradually became committed to a major role in the defence of the West, a role it has never relinquished. It is probable that the enormous worldwide economic interests of the United States would have made it adopt a more involved global policy at some point in the postwar years, but there is no doubt that the expansion of the Soviet Union in these years hastened the development as well as conditioning the form it took.

At first, the United States responded to the spread of Soviet power by economic means. In

1947, when Truman announced that he would support Greece and Turkey against Russian expansion, it was money, not arms, which was made available. The Marshall Plan, designed to help the economies of Western Europe recover, so that they could offer effective resistance to the spread of communism, was another example of this.

During the blockade of Berlin in 1948, American armed force returned to Europe in the shape of military aircraft based in Britain. However, when the NATO treaty was signed the following year, there was still little talk of a renewed arms build-up. As we have seen, American military spending was falling during these years.

Three factors combined to change this trend. First, the Russians exploded their first atomic device. This immediately brought the hint of military competition into the Cold War. Secondly, in China, Mao Zedong finally defeated the corrupt forces of Chiang Kaishek and, with his victory, another communist leader had come to power. The Americans had traditionally seen China as within their sphere of influence and for many years had supplied Chiang with aid. His downfall, though long predicted by the experts, seemed to be a major setback for American policy. Already, by this time, President Truman was considering rearming the United States to counter what many considered a worldwide expansion of communism. In the famous National Security Council Report No. 68 (often known as NSC 68) issued in April 1950, American defence experts agreed that a military response must be made. As they put it:

> In the absence of effective arms control it would appear that we have no alternative but to increase our atomic armaments as rapidly as other considerations make appropriate. In either case, it appears to be imperative to increase as rapidly as possible our general air, ground and sea strength and that of our allies to a point where we are militarily not so heavily dependent on atomic weapons.

The Korean War. A United States soldier searches for "communist" sympathisers.

In other words, not only were the number of nuclear weapons to be increased, but conventional or non-nuclear weapons were also to be built up.

It was at this point that the third factor appeared. In June 1950 the socialist state of North Korea launched an attack on South Korea. Though the North Koreans were supplied with Russian arms, it appears likely that they acted on their own initiative. The Americans decided to respond. Taking advantage of a temporary absence from the United Nations of the Soviet Union, they passed a resolution in the Security Council, which, in effect, allowed the United States and its supporters to use force in retaliation.

The Korean War was not an easy war to fight, and in fact, ended in stale-mate. However, it served to consolidate a picture in the minds of the American public. Communism was on the move. It was searching out the weak spots of the "free world", and would seek every opportunity to expand. The communists believed the victory of the revolution to be inevitable and thus they would do everything they could to help it on its way. The only possible response for the West was consistent opposition backed by military power. To achieve this, the nations of the West would have to strengthen their defences and consolidate their alliances.

This picture was oversimplified. Mao had achieved power in China without any help from Stalin. Nor is there much evidence that Stalin was behind the North Korean attack. Thus, the idea that communism was one interlinked worldwide force was much too simple. Nevertheless, this was the picture which remained clear in the American mind.

In the years that followed, the two sides became consolidated into two armed camps. NATO had been formed in 1949 and in 1955 West Germany, now a well-established independent state, was admitted to it as an equal member. The Soviet Union, in retaliation, formed its own military alliance, the Warsaw Pact, which included its Eastern European satellite states (East Germany, Hungary, Poland, Romania, Bulgaria and Czechoslovakia).

It was in these years that the arms race began in earnest. We have already seen how the United States, through NSC 68, became committed to

The Berlin Wall, the most obvious symbol of the Iron Curtain across Europe. It was built between East and West Berlin, but only after hundreds of thousands of Germans had fled from East Germany to the West.

"fighting" the Cold War from the position of military strength. The Soviet Union followed suit. Each technological advance on one side has led to renewed activity on the other. Each side has tried to maintain superiority or at least parity (equality) in weaponry with the other. We have now reached a position where each superpower is capable of destroying the other many times over.

However, it is important to consider whether the foreign policy of the Soviet Union since 1950 has been of such a nature as to justify this continued obsession with armed strength.

Crises of the Cold War

There have been three major crises between the superpowers since 1950 which deserve some examination. The first concerned, once again, Berlin. In the late 1950s the Premier of the Soviet Union, Khruschev, decided to try once more to integrate West Berlin into East Germany. He was embarrassed by West Berlin's commercial prosperity and even more by the fact that the city acted as a bolt-hole for dissatisfied East Germans. Three million of them had fled to West Berlin since 1949. West Berlin was a continuing propaganda victory for the West. Khruschev announced that West Berlin would be integrated into East Germany and then the West would have to negotiate for its future. The West stood firm, making it quite clear that they regarded West Berlin as as much a part of Western Europe as any other area. In 1961, Khruschev, in a humiliating acceptance of defeat, had to close off West Berlin with a wall, as the only way in which the East Germans could be prevented from escaping there. This move, though distressing for the East Germans, did have the positive effect of making an agreed frontier and reducing the chances of future disputes over Berlin.

The most serious conflict of the Cold War involved Cuba. Fidel Castro had come to power there in 1959, overthrowing the corrupt government of Fulgencio Batista which had enjoyed strong links with the United States. Castro gradually turned towards the Soviet Union and became increasingly anti-American. In 1961, the new American President, John Kennedy, unwisely backed an invasion force which aimed to overthrow him. The so-called Bay of Pigs invasion was a disaster and did much to consolidate Cuban suspicions of the United States and confirm what the Soviet Union already felt about its adversary. Why Khruschev now took the immense risk of attempting to provide Cuba with medium-range nuclear missiles is not clear. Perhaps he genuinely felt that Cuba needed nuclear defence against the United States. Perhaps he thought he could outclass the youthful and relatively inexperienced American President. Whatever the reason, the gamble did not pay off. The Americans spotted the missile sites and announced a naval blockade of Cuba, to prevent new military material being shipped in. The world waited tensely as the crisis mounted. Would the Russians risk breaking the blockade? Khruschev knew when he was beaten. After extracting some minor concessions from the United States, he turned his ships round and the missiles were dismantled. It was the nearest the world had yet come to nuclear war.

The third crisis came in Afghanistan in late 1979. The Soviet Union had backed a government which had been trying to build a socialist state in the highly traditional Muslim state of Afghanistan. There had been widespread opposition from the Afghans to their government and civil war had broken out. The Soviet Union was gradually drawn in, in support of the government, and in December 1979 launched a major invasion of the country. The invasion seems to have been prompted by fears that the socialist government would be over-thrown and that its collapse would spark off revolts among the Soviet Union's own Muslim peoples. The Soviet Union was forced to learn what the United States had learned in the Vietnam War. It is virtually impossible to beat guerrilla opposition which can shelter among the people and use mountainous areas for refuge. This military miscalculation was as nothing compared to the political one. The Soviet Union was roundly condemned by virtually the whole of world opinion, though it defied attempts to make it withdraw its troops.

The Soviet Union has also intervened militarily to maintain control of two of its East European satellites — Hungary in 1956 and Czechoslovakia in 1968.

Afghan guerrillas fitting up new equipment as part of their continuing struggle against the Soviet Union's invading troops.

It is difficult to summarise the Soviet Union's policy towards the Third World. Broadly speaking, it can be said that it has taken advantage of conflicts when it can, to extend its influence. It has supported liberation struggles in Africa, for instance, against minority colonial governments, and it has taken advantage of continuing Western links with South Africa to align itself to those groups who are opposing the apartheid regime there. It has built up links with several Middle Eastern states, supplying them with arms when necessary (Russia's main form of aid has been in military equipment). Trading links have also been built up with India.

A good example of the opportunism behind Soviet policy can be found in the Soviet Union's relations with Somalia and Ethiopia in the Horn of Africa. In 1974, Ethiopia was ruled by the

◄ A military superpower. The Soviet Union shows off its military strength at a parade to mark the 55th anniversary of the Revolution of October 1917.

Colonel Mengistu Haile Mariam. The Russians clearly felt that Ethiopia was a better prize than Somalia. They jettisoned Somalia and hurried in military aid to Ethiopia. When Somalia later attacked Ethiopia through the Ogaden desert, the Somalis were repelled by the arms of their former ally.

Essentially, however, it seems that the Soviet Union has maintained a defensive foreign policy. It has been concerned with maintaining its control over its satellites in Eastern Europe. Its intervention in Hungary and Czechoslovakia and its attempts to integrate Berlin into East Germany can be seen within this context. The invasion of Afghanistan, though a worrying extension of Soviet power, again seems to have been more concerned with the maintenance of control over the Muslim peoples *within* Soviet territory. The ultimate effect of the invasion may well be to strengthen the defensive nature of Soviet policy, partly because the military difficulties it has encountered may warn the Soviet Union off future military adventures and partly because of the appalling image the invasion presented to the rest of the world, particularly to the non-aligned nations, among whom the Soviet Union wishes to extend its influence.

The attempt by Khruschev to supply Cuba with missiles appears at variance with this policy. It was certainly the most directly provocative move made by the Soviet Union since 1950 and remains to be fully explained. What it does suggest, however, is that the Soviet Union is prepared to extend its influence if it feels it will be able to do so without firm reaction. In this, the Soviet Union is not far different from most militarily powerful states throughout history.

It is, in fact, the military strength of the Soviet Union which must remain the most worrying feature of its policy. Over the past few years the Soviet Union has committed between 11 and 15 percent (Western estimates vary) of its national income to defence, and its spending seems to be continuing to rise. Soviet forces, both nuclear and conventional, have been modernized and

aging and highly conservative Emperor Haile Selassie, who had strong military links with the United States. Ethiopia's neighbour, Somalia, had a long-standing dispute with Ethiopia over their common border, and the Soviet Union took advantage of this to align itself with the left-wing government of Siad Barre in Somalia. In 1974, however, Haile Selassie was overthrown and a pro-Soviet military government emerged, under expanded and the country has developed a strong naval capability.

There have been numerous attempts to measure the strength of Soviet military power against that of the members of NATO, but it is virtually impossible to come to an accurate assessment. It is not just numbers of weapons, tanks and ships which matter, but their efficiency at times of war, the skill of those operating them and the morale of the fighting forces. What is clear is that there has been an important build-up in Soviet forces and that these would offer an overwhelming threat if war broke out. In some areas, such as tanks, artillery and certain types of aircraft, the Warsaw Pact nations considerably outnumber the forces of NATO.

The Soviet economy has developed major weaknesses in recent years and thus the continuing commitment of such large resources to defence spending is worrying. How can we explain this commitment?

One reason for it could be a continuing Soviet obsession with security, in what it perceives as a threatening world. As we have seen, Stalin's policy seems to have been largely conditioned by such fears, both as a result of the enormous suffering the Soviet Union endured in the Second World War and because of what was seen as the enduring hostility of the capitalist nations. Since 1950 the Soviet Union has been faced by a wealthy United States, determined to maintain a superiority in weapons and technology. In 1959 there was a major rift between the Soviet Union and China. The hostility between the two countries became as bitter as any between the Soviet Union and the West, and intensified as China started to build up links with the West from the 1970s. With the NATO members on one side and a hostile China on the other, the Soviet Union can point to a genuine fear of being encircled.

It would seem that this desire for security, with the only real security being seen in military strength, is certainly part of an explanation for the Russian build-up of arms. It may also be that the Soviet Union sees military strength, not just in terms of the power it brings, but as the *symbol* of being a superpower. The Soviet Union has always needed to show that it can equal the United States in achievement, and maintaining a wide range of modern and sophisticated weaponry is one way it can do so. It is interesting to note that the Soviet Union regards military aid as the best way of building up influence in the Third World. According to a recent report by SIPRI:

> the Soviet Union uses arms transfers as an important instrument for maintaining and expanding its influence in the Third World. Arms play a greater role in Soviet relations with the Third World than economic aid or trade.

Some have continued to argue that the only realistic explanation for the build-up of arms is that the Soviet Union has plans for renewed aggressive expansion. It is difficult to find much evidence for this. As even the relatively limited incursion into Afghanistan has shown, military power is a blunt instrument and an increasingly difficult one to use with much political effectiveness. The evidence suggests that the present Soviet leadership and its likely successors are cautious by nature. They will face considerable problems in keeping the Soviet Empire together, let alone in expanding it.

Nothing in what we have said rules out the possibility that a sudden crisis might blow up, in which both sides find themselves locked in conflict. What chances are there of this?

The original conflict of the 1940s over the extent of Soviet influence and control in Europe appears largely to have been solved. In the final act of the Helsinki Agreement of 1975, both East and West accepted as permanent the borders in Europe that had been established after the Second World War. As we have seen, the status of Berlin now seems accepted by the Soviet Union as well.

However, there are other areas of potential conflict. The Middle East continues to offer the prospect of endless instability. It has important oil reserves and either superpower might be drawn into conflict there, in a search to extend or maintain influence. The other might then respond and a crisis become under way.

It is just possible that the West might be tempted to intervene in support of an Eastern European country trying to break free of Soviet control. As we have seen, this area is seen as vital to Soviet defence and Western intervention would, without doubt, provoke a reaction. Another possible crisis point is China. At present, the Soviet Union is much stronger than China. If China continued to consolidate its links with the West, there might be a moment when the Soviet Union decided to try to defeat it. Here again, the West might be drawn in, in response.

If a crisis did occur, what chances are there that it would be defused before it reached the level of nuclear war? It is certainly true that the awesome risks presented by the possession of nuclear weapons have encouraged the two superpowers to talk of means of crisis management. In the early 1970s, for instance, when relationships between East and West improved for some years, a number of agreements were made. In the Declaration of Basic Principles of Relations of May 1972, it was stated:

> the USA and USSR attach major importance to preventing the development of situations capable of causing a dangerous exacerbation of their relations. Therefore, they will do their utmost to avoid military confrontations and to prevent the outbreak of nuclear war. They will always exercise restraint in their mutual relations and will be prepared to negotiate and settle differences by peaceful means.

Since these agreements, tensions between the superpowers have increased, following the Soviet invasion of Afghanistan in 1979 and the emergence of the more conservative and strongly anti-soviet government of President Reagan in the United States. It remains to be seen how strong the agreements would be in time of crisis. The influential International Institute for Strategic Studies is obviously not too impressed with them. In its Strategic Survey for 1981-82 its assessment was that "procedures for effective crisis management remained woefully undeveloped".

Many defence experts claim that the real threat of nuclear war comes from the development of tactical nuclear weapons which would make the resort to nuclear war that much more easy in times of crisis. This point will be explored on page 54.

The Soviet Threat — a Summing Up

It does not seem that, at the present time, the Soviet Union is committed to a policy of overseas expansion. It is preoccupied with severe economic problems and has great difficulty in maintaining its influence over its own Empire and the nations of Eastern Europe. Its leadership appears conservative by nature. This conservatism should continue as long as the Soviet Union sees no obvious advantage from expansion. The risks of expansion remain high, not only in terms of the military response possible from the West, but also in terms of the political damage any expansion could do to the Soviet Union's image in the world community.

The most important threat to peace would seem to lie in some unforeseen crisis arising in an area of instability, where both superpowers, or their allies, felt they had vital national interests to protect. The situation would be especially dangerous if one side miscalculated the importance of the issue to the other. (This is what happened in the Cuban missile crisis of 1962.) In such a case, if the procedures for crisis management broke down, then the risk of violence, escalating perhaps to the use of nuclear weapons, would be grave.

The implications for Western defence policy would seem to be as follows. First, while the Soviet Union appears to offer at present little immediate threat of expansion, this is partly because it knows full well the risk of doing so. In present circumstances, it would seem prudent for the West to maintain sufficient forces so that the Soviet Union is not tempted to reassess its policy.

Secondly, renewed efforts should be made to improve crisis management. The greatest risk of war, it would appear, comes not from planned aggression by either side but from miscalculation. With obvious and increasing global instability, the need to manage crises skilfully when they arise is paramount.

Thirdly, the vast arsenals built up by each of the superpowers and their allies make the possibility of war a terrifying one. Some have argued — and, in fact, evidence can be found to support the argument — that the possession of nuclear weapons has made the superpowers act more responsibly at times of tension between them. However true this has been, the evidence is certainly not strong enough to suggest this as a legitimate policy for the future.

THE BRITISH RESPONSE

In this chapter we are going to examine the development of British defence policy since the Second World War. Broadly speaking, there are three main themes which we can distinguish. First, over these years, Britain has made a major adaption from conducting what was a global defence policy to one which is overwhelmingly concentrated on Europe. Secondly, for the first time in its history, Britain has been a member of a long-term and relatively stable military alliance, NATO. Third, Britain has placed important reliance on independent possession of nuclear weapons, maintaining these as a deterrent against attack. It is this last issue which has proved most controversial, and the one which we will examine in most detail.

The Costs of Defence

The most important influence on the development of British defence policy has been money. The Western democracies are not warlike by nature, and after the experience of two major wars, there has been a deep revulsion against any glorification of war. It is in the West that there have arisen the largest and most active peace movements. There has been very little enthusiasm for maintaining heavy expenditure on defence, and it has remained a low priority in the public spending of most of these nations. In Britain, for instance, the percentage of national income spent on defence gradually declined until the Conservative government elected in 1979 promised to increase it.

Here are some figures from SIPRI for British defence spending as a percentage of national income:

1955	8.2%	1976	4.9%
1960	6.5%	1979	4.7%
1965	5.9%	1981	5.0%
1970	4.9%		

Until the recent increase, the decline has been steady and consistent. The decline is even more obvious when we look at the share taken by defence in public spending (government spending) as a whole. While, in 1953, the British government spent 28.5 per cent of its total spending on defence, by the 1970s the figure was down to 11 per cent. In these years, other services, such as the health services, social security and education, had all overtaken defence. It is important to make this point because there is a common belief that expenditure on defence is escalating fast. This would certainly seem true of many Third World nations, but it is not typical of Western Europe.

On the other hand, within Europe, Britain devotes a higher percentage of its income to defence than do most of its allies. Again according to SIPRI figures, Germany spent 3.4 per cent, France 4.2 per cent, Belgium and Holland 3.3 per cent and 3.4 per cent respectively of their national incomes on defence in 1981. The United States spent 5.8 per cent.

Although the percentage of its national income spent on defence has fallen over the past thirty years, Britain has become more wealthy as a country and the actual amount of money available for defence has risen slightly. This has been more than offset by the ever increasing costs of new equipment. In fact, Britain has been able to obtain less and less for its money every year and, as a result, its defence commitments have had to be steadily cut.

While lack of money has been the major force influencing defence policy, there have been other

Britain still has great pride in its navy. How far it can continue to maintain a strong navy when costs are soaring remains to be seen. Here is the Aircraft Carrier *Invincible* which played a prominent part in the Falklands Task Force, 1982.

important factors. One is tradition. Britain has traditionally maintained a large navy and a small standing army. Its navy remains the third largest in the world, after those of the United States and the Soviet Union, and there are strong, if not always successful, pressures to maintain it as the backbone of the British defence commitment. Another factor, and one which has been central to British defence policy, has been commitment to allies through the NATO pact. This has strongly influenced the types of forces kept available and their positioning. The maintenance of a British army in Germany is an example of this.

The British defence programme has also been conditioned by what technology is available. Weapons systems are so expensive that even the richest nations are limited in what they can produce. Britain has not been able to afford to develop much technology of its own and is often forced to accept what the United States happens to have on offer. This may mean weapons developed for a completely different purpose than that required by Britain. An example is the Trident missile, which is unsuitable for Britain's needs in several ways, but is claimed to be the only possible alternative to Polaris (see page 49 for a discussion of this issue).

The Retreat to Europe

As Britain gave independence to its Empire in the twenty-five years following the Second World War, its immediate responsibility for the Empire's defence disappeared. Yet both major political parties in Britain continued to insist that a defence role be maintained outside Europe, in particular east of the Suez Canal, both in the Middle East and further east in South-East Asia. Britain had important economic interests in each area; rubber and tin in the Far East, oil in the Middle East. Both areas were unsettled. Britain's decision to maintain military bases east of Suez (these included Aden, Ceylon, Singapore and Hong Kong) was supported by the United States, which welcomed a continuing British presence in what were potential areas of conflict.

However, by the 1960s, Britain's weakening economic position made it obvious that an east of Suez role could not be maintained. After the major economic crisis of 1967-68, the Labour government then in power accepted that the withdrawal had to come and troops left, finally, in 1971. The Conservative Party which was far more deeply and emotionally committed to a global defence policy howled its protest at the decision, but did nothing to reverse it when it next came to power. However painful a decision it was for some, it was without doubt the right one.

Britain's defence efforts thus became concentrated in Europe, with the main emphasis on Britain's role within NATO. As we have seen, NATO was founded in 1949, as a response to the expansion of the Soviet Union. It marked an important step in European cooperation. Never before had a number of nations formed a long-term alliance of this kind. NATO was equally important in integrating the United States and, to a far lesser extent, Canada into the defence of Europe.

Britain has maintained especially strong links with the United States. The two nations share a common language and many elements of the same culture. They fought as allies in both World Wars, with the relationship becoming especially close in the Second. Their common membership of NATO has served to consolidate the relationship further and, as we shall see, Britain has relied heavily on the United States for its nuclear weaponry. Britain, like the other European members of NATO, benefits from the United States' bearing a very high proportion of costs of the alliance (nearly two-thirds of the total).

One side-effect of this close relationship has been that Britain has served as a base for the troops and weapons of the United States. For instance, there are 170 F-111 nuclear bombers stationed in Britain. The most controversial of these weapons have become the Tomahawk Cruise missiles, of which 96 will be based at Greenham Common near Newbury by the end of 1983 and a further 64 at Molesworth in Cambridgeshire in 1988. These weapons are to be introduced in response to the SS-20, a medium-range missile which the Soviet Union has targeted at Western Europe.

Cruise missiles are pilotless aircraft designed to fly at less than 150 feet, following mapped-out computer routes across Europe. The Cruise missile carries one small nuclear warhead and

The Cruise missile. These pilotless aircraft fly amost at ground level so as to avoid radar detection. Britain and other NATO members have agreed to deploy them, a decision which has aroused much protest from peace campaigners.

therefore is not one of the most destructive nuclear weapons. It is designed to be dispersed in time of crisis, so that it would not be vulnerable to attack from the Soviet Union. However, since the missile bases are well-known, there could be a temptation for the Soviet Union to attempt a "first-strike" in order to knock them out.

The Cruise missile has aroused a torrent of criticism, and the decision to base Cruise in Europe has been as responsible as any other for reviving renewed protests against the whole policy of nuclear defence. There are fears that the missile bases would make Britain more vulnerable to a first-strike nuclear attack at a time of crisis. These are also fears that, since the Cruise is a less destructive weapon, it might be seen as little different from conventional weapons and thus there might be little inhibition about launching it at a time of military conflict. It has been forcefully argued that its use would set off the process by which escalation to the strategic nuclear weapons with their devastating destructive power would take place. Fundamental to all these protests is a refusal to believe that any more weapons of destruction could possibly be needed when the world can already be destroyed several times over. Arguments of the nuclear strategists that every level of Soviet weaponry should be met by an equal level of Western response no longer seem convincing.

We will be examining the whole concept of nuclear deterrence in the next chapter. Before looking at Britain's own independent deterrent, it is important to say something about the role of conventional forces provided by Britain to the NATO alliance.

Britain maintains two "conventional" roles within NATO. The first is to provide army units based on the Rhine, a total of 55,000 men. These are supported by air force units. Many of the British aircraft have what is known as "dual-capacity"; they are able to carry either conventional bombs or nuclear weaponry. The maintenance of this force permanently on the European mainland marks an important change in Britain's traditional defence policy. In both World Wars there was no British presence on the mainland until troops were hurriedly gathered at the outbreak of war and then rushed over. It is fair to say that Britain's commitment to the frontline defence of Germany remains the most important symbol of its political commitment to the defence of Europe.

It might equally be argued that Britain's most effective commitment lies with its navy forces. A strong navy has, of course, been central to Britain's traditional defence policy. The British navy, despite many cutbacks and economies, is still large, the third biggest in the world. In time of war, its role would be partly to contain any Soviet naval forces, in particular those attempting to break through the "gaps" between Britain, Greenland and Iceland. Its other role would be to defend shipping bringing in supplies from outside Europe.

It is worth repeating a point made earlier.

Britain can withdraw its troops from NATO at any time it wants. Its army commitments in Northern Ireland and the launching of the naval task force in the Falkland Islands crisis have both been possible, without breaking any of the terms of the alliance. It is, of course, politically wise for Britain to maintain its commitments to NATO as effectively as its economy allows.

In the 1980s, it would appear that Britain will have to make the decision whether its commitments to NATO should be primarily naval or military on the European mainland. It is becoming clear that the increasing expense and sophistication of modern weaponry are likely to make it impossible for Britain to maintain both roles effectively. The 1981 Defence White Paper (the statement on the development of Britain's defence programme, published each year) argued that the land forces are more important and that the British naval forces should be once again cut. Whether the Falkland Islands crisis would affect this decision remained to be seen.

Britain's membership of NATO has been overwhelmingly accepted by the vast majority of British people and politicians. The advantages of sharing the burdens of defence are obvious and the independent efforts of even the wealthier European nations would look tiny compared to the forces of the Soviet Union. A public opinion poll, commissioned by the BBC in 1981, showed that only 9 per cent of the British public wished to withdraw from NATO. Two groups, however, do continue to argue for a restoration of Britain's traditional independent role. The first is a conservative group, who are worried about Britain's continued dependence on the United States and wish the country to be totally in control of its own defence policy. Their most prominent spokesman has been Alan Clark, the Conservative

A Rapier missile air defence battery — as used by British forces in the Falklands. Concern over nuclear weapons must not blind us to the growing destructive power of conventional weapons.

MP for Plymouth Sutton. He has argued that NATO has proved very ineffective as a coordinating body for defence policy and that Britain should leave and formulate its own. He has talked of a "Fortress Britain", which would be able to resist all attackers, with a continued reliance on nuclear weapons to deter aggression. At the other political extreme are those who argue that it is the combined forces of the NATO alliance which are most responsible for the continued build-up of Soviet strength (in self-defence) and which thus form a major cause of world tension. If NATO could be dissolved, the Soviet Union would relax its own defence efforts and world tension would be reduced. A first step would be for Britain to withdraw unilaterally. Whether Britain, occupying as it does an important strategic position in Western Europe and with so many political, economic and ideological links with the Continent, could revert to this independence is unlikely. Some "decoupling" of the European members of NATO from the United States might be more possible.

Since 1945 no conflict in the world has involved the use of nuclear weapons and it is clear that conventional weapons are still considered the most effective way of achieving military objectives. As the British recapture of the Falkland Islands in 1982 showed, conventional weapons and human skills still have a vital role to play in modern warfare. It is important to remember that conventional weapons can also be extremely destructive, and there is now a considerable overlap between the destructive force of the more advanced conventional weapons and that of the less powerful "tactical" nuclear weapons.

The British Nuclear Deterrent

An important feature of British defence policy since the 1950s has been, however, a reliance on an independent nuclear force as a deterrent against attack. The decision to develop British nuclear weapons was made secretly by the British Labour government in 1947, but it was not until a Conservative government came to power, in 1951, that the programme was made an important element of the British defence system. The first British bomb was exploded in October 1952, making Britain the world's third nuclear power (after the United States and the Soviet Union), and a thermonuclear or hydrogen bomb was tested in 1957. Right from the start, the deterrence nature of the bomb was stressed, in the belief, in the words of Lawrence Freedman, that "the destructive nature of nuclear weapons was considered so great that no nation would dare to provoke a war in which there was the slightest risk of it becoming the victim of an attack using them". This deterrent role, which will be examined more critically in the next chapter, has remained the major one for these weapons. Why Britain has needed its *own* nuclear forces beyond those provided by the United States is not so clear. It was argued in the 1950s that they would help maintain Britain's position as a great power, but there is little evidence that this has been the case. An argument put forward by the French, who also have independent nuclear forces, is that the long-term future of the NATO alliance, in particular the continuing commitment of the United States, cannot be guaranteed, so that it remains important for nations to retain an independent nuclear force. The British have been rather reluctant to follow this argument, for fear that it may appear to the United States that they do not totally trust American commitment to Europe.

The official reason given for the retention by Britain of an independent nuclear deterrent is that it adds another centre of decision-making at a time of crisis. If the Soviet Union were considering an attack, it would have to assess the reaction not only in the United States but also in Britain and France. The total deterrent value of three centres, in which the decision whether to launch nuclear retaliation would have to be made, would, it is argued, be greater than just one.

Such an argument has been hotly criticized, not least by a former Chief of the Defence Staff, Field-Marshall Lord Carver, who has argued that it is impossible to conceive of a situation where Britain would want to launch nuclear weapons when the United States would not. It could also be argued that, at a time of international crisis, when emotions, fears and suspicions are rising high, to have three centres of decision-making would be far more likely to increase the instability of the situation.

In short, it is difficult to find any overwhelming argument for the maintenance of an independent deterrent. If Britain did renounce its deterrent, it would still have to make the decision whether to continue to rely on the United States' nuclear weapons. As we have seen (page 22), there are dangers in relying on a nation which has such different perceptions of world politics. It might be that there is a case for a European-controlled nuclear deterrent. This cannot mask the problem which we will examine in the next chapter, that *any* policy relying on nuclear weapons for its ultimate defence contains serious dangers for the future of mankind.

An important reason for Britain's embarking on a nuclear defence programme was the cheapness of such weapons, in comparison to equivalent forces of conventional weapons. In 1957, the Defence Secretary of the day, Duncan Sandys, issued his "Outline of Future Policy". He argued that the possession by Britain of a nuclear deterrent reduced the need for expensive conventional forces. Thus, these could be run down, with the end of the conscription of forces. This has been one more area where cost has been the most important factor in deciding the form of defence forces.

It was one thing to decide that Britain should have its own deterrent; it was another to find the right weapon. At first, Britain, like the other nuclear nations, envisaged dropping its bombs from aircraft, but by the late 1950s it was clear that missiles had replaced planes as the most effective method of delivery. Britain decided to develop its own missile, Blue Streak. But it soon became clear that this would be obsolete even before it was completed, and thus Britain had to fall back on collaboration with the United States (which had, in fact, encouraged the development of a British nuclear programme). The Americans offered their new missile, Skybolt, and the British government placed all their hopes on acquiring this. In 1962, the news reached London that the Skybolt programme was being cancelled. It was simply too expensive and had major technological problems.

The cancellation of Skybolt was a shock to Britain. It brought home to everyone just how dependent Britain's nuclear effort was on what

The Polaris missile. Here one of the first missiles, developed in the early 1960s, is fired in test. Polaris remains the major element of the British nuclear programme and will continue to do so until the 1990s.

the United States happened to have on offer. To the relief of those who had staked the future of British defence policy on the possession of a nuclear deterrent, there was an available alternative — the Polaris submarine-launched missile. Britain was able to do a deal with the United States, by which it obtained Polaris for a very reasonable sum. The United States would provide the missiles, Britain the warheads and the submarines in which they would be placed.

Polaris is a missile with a range of some 2,500 miles. It is relatively cheap; the total Polaris force, even with money spent on research and development, has cost on average only 4 per cent of the total defence budget. It is at present invulnerable to attack when the submarines are at sea (and thus it is virtually impossible to knock out in a "first-strike"). It also has a reasonably long life. The Polaris force, with the renovations mentioned below, will last well into the 1990s (in other words for a total period of thirty years) and, some have argued, for even longer.

There have had to be renovations and refittings, however, and these have proved expensive. By the early 1970s there were discussions as to whether the quality of the British weapons should be upgraded. One possibility was to introduce a completely new missile, the Poseidon. However, it was finally decided to replace only the warhead of Polaris. The new warhead was called Chevaline, which, it was claimed, would be better able to penetrate Soviet defences. The decision to go ahead was made in 1974 by the then Labour government; the estimated cost was £200 million. It was a secret decision, only officially announced in 1980. By this time, the cost of Chevaline had soared to £1,000 million. In 1982 it was also announced that several hundred million pounds would have to be spent on new motors for the Polaris missile.

It was clear from these events that the costs of major defence programmes of this nature are simply not predictable. This is typical, of course, of all programmes involving major technological innovation. Equally serious was the realization of how easy it was to conceal major sums of expenditure from the public. The Chevaline programme was under way for six years before the large costs involved were made known. This does mean that one element of defence policy, the spread of expenditure, has taken place without public scrutiny. Altogether, there has been increasing public criticism of the way the Ministry of Defence has managed its expenditure.

The expensive renovations mentioned above have been followed by another major decision. The question was, what would happen in the 1990s, when even the updated Polaris became obsolete? The Conservative government announced in 1982 that Britain was to replace Polaris with another submarine-launched missile system, the Trident. The first model of the Trident discussed, the C-4, was a missile with a range of 4,500 miles and with eight warheads on

A "Polaris" submarine, in this case one belonging to the United States, visiting the nuclear base at Holy Loch, Scotland.

each missile. (Polaris has three.) This model is being used by the United States navy, but is to be phased out in the late 1980s, with the American production line being closed in 1985. The United States' replacement is a bigger Trident, the D-5. This has a range of 6,000 miles and there are fourteen warheads on each missile. Each one is independently targeted and is claimed to be extremely accurate. The British government announced that this would be the version they would buy.

The Trident D-5 will be much more expensive than the C-4. For a start, it is larger, and therefore will need a larger submarine to be designed for it. Estimates made in 1982 were that the full replacement of Polaris by Trident over the next fifteen years would cost between seven and eight billion pounds. However, as we have seen with the Chevaline programme, costs of missile technology are notoriously difficult to predict and the final cost may be very much more.

Cost is not the only problem associated with the Trident. There is some argument for having a missile with increased range, as the most import-

The Campaign for Nuclear Disarmament on the march. This was the first of the famous marches from London to the nuclear weapons research station at Aldermaston, at Easter 1958.

ant feature of a deterrent is that it should be invulnerable to attack. The larger the area in which a submarine can conceal itself, the more invulnerable it will remain. However, transferring to Trident will mean that Britain will now be defended by 640 nuclear warheads as compared to 192 at present. This is a breach of an agreement which the British government has signed (the Non-Proliferation Treaty of 1970) not to increase the number of warheads in this way. Also, it is not certain that Britain needs a missile of such accuracy.

When the British nuclear programme began in the 1950s, there was widespread opposition to it, coordinated by the Campaign for Nuclear Disarmament (CND). In the 1960s and 1970s CND's support began to wane and it appeared that people were beginning to learn to live with nuclear weapons. The Trident decision, as well as the decision to base Cruise missiles in Britain (see page 44) have helped to revive a groundswell of opposition to nuclear weapons.

The opposition to Trident can be divided into three groups. First, there are those who wish to keep Britain's independent deterrent, but feel that Trident is the wrong choice. Other options they put forward are a submarine-launched Cruise missile or some form of updated Polaris which would not be as efficient as Trident but which would still present enough risk to act as a deterrent.

Secondly, there are those who, while continuing to believe in nuclear deterrence, wonder whether the time has not come for Britain to give up its *independent* deterrent. They would argue that the value of the independent deterrent is not large when put alongside the other nuclear forces of the NATO alliance and opportunities for wishing to use it independently are not evident. It would be better to spend the £8,000 million on maintaining the strength of Britain's conventional forces. Britain would continue to rely on the United States' nuclear weapons.

Third, there are those who reject all forms of nuclear weapons, not only Polaris and Trident, but also the NATO nuclear deterrent provided by the United States. We will examine the views of this group in the next chapter.

Trident is a long-term programme. The building of submarines and the transition from Polaris will take several years of continued expenditure. It will be important, therefore, that all British political parties support the decision. At present, only the Conservative Party is committed to the Trident decision and thus the programme may well be subject to cancellation if a change of government takes place.

At present, Britain's independent nuclear deterrent does appear to be the least justifiable of its defence programmes. It may well be that in the 1980s the decision will be made to cancel it and to spend the money saved on conventional armed forces.

In this chapter we have discussed British defence policy in terms of Britain's own commitments and alliances. In the next chapter, we need to look in more detail at the implications these, and the policies of the superpowers in general, have had for the long-term security of the world.

WHAT'S WRONG WITH DEFENCE POLICY?

As we have seen, there are certain fundamental questions which each nation has to answer when formulating a rational defence policy. First, what is the nation defending? What national interests are so important that we ought to be prepared, in the last resort, to use armed force to defend them? Secondly, what threats are there to these interests? As we have seen in the chapters on "Threat Assessment" and "A Soviet Threat?", the assessment of threats is enormously difficult. The miscalculation of potential threats, in fact, remains one of the most common causes of war. The third question is perhaps the hardest to answer. Accepting that a threat is seen, what is the best way of defence against it?

The Western alliance has taken as the central element of its defence policy the concept of deterrence — that is, the threatened retaliation by nuclear weapons, against any aggressor, in the hope of deterring the aggression. It goes without saying that the use of the threat of nuclear weapons in itself causes immediate concern. The so-called strategic nuclear weapons continue to be largely targeted at cities and to contain hideous destructive power. Furthermore, we have still all too little experience of how to handle nuclear weaponry within the context of possible warfare.

Lawrence Freedman, who has written what is by far the best analysis of nuclear strategy, puts the problem as follows:

> We do not know how political leaders will react to the news of even only a few nuclear weapons exploding on their territory. The response could be a reckless fury, lethargic submission, craven cowardice, or a firm and resolute action. We do not know how an American President will react to the news that Soviet forces are advancing through Western Europe. He may be struck most by the danger to his country's population if any steps are taken towards nuclear war, or by a sense of obligation in honouring alliance commitments. Not only do we not know how our leaders will respond: still less do we know the expectations of Soviet leaders as to Western responses or how they themselves would cope with their own moments of truth: and even less do we know how things will have changed in a decade or so when the weapons now being conceived will become operational. (L. Freedman, *The Evolution of Nuclear Strategy*, Macmillan 1981)

The Doctrine of Deterrence

The concept of deterrence, on which the security of the Western alliance is supposed to depend, is fairly straightforward, and is not a new one in the history of warfare. It is based on the belief that, if you make the risks of an attack on your territory far higher than any possible gains which an attacker may make, then he will be discouraged from attacking. In our present deterrence strategies, the "risks" are provided by nuclear weapons, capable of causing enormous damage to the aggressor's territory. The main targets would be his cities.

For deterrence to be a credible policy, it must contain the following elements. First, you must possess sufficient armed power, whether in nuclear or conventional weapons, to carry out significant damage to the aggressor. If nuclear missiles are the deterrent, they must be capable of breaking through any enemy defences. Secondly, the enemy must believe that you will actually use your weapons at a time of attack. Third, your weapons must be invulnerable to a "first-strike"

CND reborn. Following some years of decline, CND saw a major revival of support in the early 1980s, as decisions were made to bring Cruise and Trident missiles to Europe. Here is the mammoth CND demonstration in London in October 1981.

attack. If an enemy knows he can knock out your deterrent weapons, he could be tempted to do so in a "first-strike" — and your deterrent could have, in effect, caused, rather than prevented, war. In short, the theory of deterrence is that, if you have a deterrent force which is invulnerable to attack and which your enemy believes you will use in time of war, he will be deterred from attacking.

As nuclear weapons have such enormous destructive power, the theory of deterrence would suggest that all that is needed is a relatively small stock of them. When we look at the tens of thousands of warheads that the superpowers, in fact, have ranged at each other, the theory seems to have become a nonsense.

It is not easy to explain why such large numbers of weapons have been built up. At first, there appeared to be a continuing belief, lingering on from pre-nuclear days, that numbers of weapons, or total destructive power, were a real measure of military strength, even though this destructive power was many times greater than any conceivable use that could be found for it. It was only in the 1970s that the superpowers seemed to be growing out of this belief.

The arms race has also been fuelled by the continuing suspicions between the superpowers. Each move by one side is interpreted by the other at its most sinister. If we read statements by leaders of the superpowers, we continually come across the claim that, while they are acting purely defensively in, say, developing a new missile, a similar move by their opponent is evidence of possible aggression. So, it is argued, some form of response is necessary. The total number of weapons increases. The problem is made worse by the fact that each superpower has tended to overestimate the other's strength, and has used this overestimation as the basis of a call for yet more arms.

Technological developments have also been responsible for the build-up of numbers. If we look at Britain's own deterrent, for instance, we find that, with the acquiring of Trident, the total number of missiles deployed by Britain will jump from the 192 of Polaris to 640. There is no reason for this increase other than that the Trident system is considered by the British government as the only suitable replacement available.

Technology has thus conditioned a massive increase in nuclear power.

By the 1970s the superpowers began to realize the absurdity of the situation that had evolved and they met together to agree on ceilings to the numbers of strategic (that is intercontinental) weapons they would deploy. The first SALT (the initials stand for Strategic Arms Limitation Talks) agreement came in 1972, when both sides agreed to limit the number of missile launchers they had, both on land and at sea in submarines. The SALT II agreement was signed in 1979 and limited both the United States and the Soviet Union in the numbers of various types of strategic weapons they could deploy. The United States has not formally ratified this agreement, but has, in fact, observed it.

The agreements are what we call arms control agreements. They should prevent further increase in numbers of strategic weapons, but, for the most part, they do not include any measure of disarmament. So, while the SALT agreements may be taken as evidence that both superpowers are concerned about the arms race and prepared to put some limitations on it, both nations remain as heavily armed as before. It is even possible that they will put more effort into building up stocks of weapons which are not covered by the agreements.

Tactical Nuclear Weapons

One area where there is still unrestricted build-up of arms is tactical or theatre nuclear weapons. These are the shorter-ranged weapons based in Europe. On the whole, they are less destructive than strategic nuclear weapons and many of them are aimed at military targets rather than cities. (It must be remembered, however, that many cities are military targets and that the destructive power of these weapons would still make civilians major casualties of their use.) There are two main reasons why the Western powers have placed a continuing reliance on these tactical weapons. First, it appears that the conventional forces of the Soviet Union and its allies in Europe outnumber those of the NATO members (there is naturally some dispute as to how far this is true, but most experts accept that it is the case). To make up for its shortage of conventional forces, the West believes that the greater destructive power of tactical nuclear weapons is needed, to keep some form of balance. The second reason is to maintain some credibility for the deterrence forces. In the early days of deterrence, when the United States had an overwhelming superiority in nuclear weapons over the Soviet Union, it developed the policy of "massive retaliation". As the American Secretary of State in the 1950s, John Foster Dulles, put it, the United States was to "depend primarily upon a great capacity to retaliate, instantly, by means and at places of our own choosing" with nuclear weapons against any Soviet attack. There were two drawbacks to this policy. First, the Soviet Union in the late 1950s also developed large numbers of strategic nuclear weapons, so that massive retaliation would be met by massive retaliation, leaving neither side the winner. Secondly, it was difficult to envisage the circumstances when the full horrors of massive retaliation might be unleashed. The effects would be so disastrous that the United States might well refrain from using any nuclear response at all to minor cases of Soviet aggression.

It was argued that the development of tactical nuclear weapons would help solve this problem. Because they were less destructive, they could be used in a much more limited way, to deter minor acts of Soviet expansion, in cases where full massive retaliation would not be appropriate. By being used against a few isolated Soviet targets, they could act as a warning of what might happen if aggression continued. Altogether, they provided a deterrent force which was much more credible than one relying just on massive retaliation.

The acceptance of tactical nuclear weapons led to the doctrine of so-called "flexible response" adopted by NATO in the mid-1960s. Laurence Martin explained the doctrine in his recent Reith Lectures on "Armed Force in the Modern World":

> The first line of Allied resistance to conventional attack is to be conventional resistance; if that fails, tactical nuclear weapons used on the battlefield are both to reinforce the defence and raise the spectre of escalation; finally, if all else fails, weapons are to be employed more widely and ultimately 'strategically' against the Soviet Union.

The continuing development of tactical nuclear weapons, particularly the introduction of new missiles by both sides (the SS-20 by the Soviet Union and the Cruise missile by the West), has revived a debate, which first took place in the 1950s, on how far these weapons increase the risk of nuclear war. One of the constraints on the use of nuclear weapons has been an awareness of their enormous destructive power. The less destructive nuclear weapons become, the more imaginable becomes their use. The less destructive and more refined tactical weapons may be seen as little different from a conventional weapon. The order to use them could be given that much more easily in time of crisis. Some fuel for these fears was given by the issue in 1980 of the so-called Presidential Directive 59 by the US President, Jimmy Carter. The directive announced that, instead of the United States retaliating at a time of war by launching nuclear missiles at Soviet cities, it would now aim at military targets, command centres, missile sites and so on. At first sight, this may appear to be reassuring, but there are three reasons why it is not. First, as we have argued above, it makes the use of nuclear weapons, early in a conflict, more imaginable. Secondly, even military targets are often surrounded by civilian populations. Command centres are often based in capital cities, for instance. Even with the increased accuracy of nuclear weaponry, many weapons would still miss their targets and presumably hit civilian populations. Third, once the nuclear weapons are used at tactical level, it is highly likely that there will be, in any case, an escalation to the full use of strategic weapons.

Many strategists have discussed this last point and there seems virtually complete agreement that a nuclear war, once started with tactical nuclear weapons, could not be prevented from escalating. As Lord Mountbatten said in a much-quoted speech made a few months before his murder:

> The belief that nuclear weapons could be used in field warfare without triggering an all-out nuclear exchange leading to the final holocaust is more and more incredible. I cannot accept the reasons for the belief that any class of

nuclear weapons can be categorised in terms of their tactical or strategic purposes. In all sincerity, as a military man, I can see no use for any nuclear weapons which would not end in escalation with consequences no one can conceive.

Criticisms of Deterrence

The first criticism that can be made of the policy of deterrence is, therefore, that it has not led to a stable balance of power between the superpowers, based on a limited but effective threat of nuclear retaliation in case of aggression. Suspicions, technological development and an obsession with numbers have continued to keep the relationship between the powers an unstable one.

Many have argued that these factors have made nuclear war more likely in the 1980s than it has ever been before. Of course, such a statement cannot be proved either way and we can only speculate whether this is so. It is equally possible to argue, as the International Institute for Strategic Studies has done in its Strategic Survey for 1981-82, that the relative balance between the nuclear forces of the two superpowers has, in fact, made war less likely compared with in the 1950s and early 1960s, when the United States had overwhelming superiority in weaponry and could have possibly "won" a nuclear war. The fact remains that, with the large number of weapons deployed by each side in an atmosphere of continuing suspicion, there can be no great cause for optimism about the future of mankind at a time of crisis.

Several other criticisms can be made of the doctrine of deterrence. Deterrence is needed because you are suspicious of the intentions of a potential enemy. By targeting nuclear weapons at him, you are simply reinforcing mutual suspicion. This may make it far harder to find constructive political solutions to the problems on which the initial suspicions were based. By sheltering behind their nuclear weapons, the two superpowers are restricting their chances of reaching permanent solutions to their differences.

Protest worldwide — another anti-nuclear demonstration, this time in New York.

For deterrence to work — in other words, for deterrence to be a policy whose primary aim is to avoid war — it must be absolutely clear to both sides exactly what actions are supposed to be deterred, so that the potential aggressor knows exactly what he must avoid doing. In the confrontation between East and West, it seems clear that an attempted invasion of either East or Western Europe by the opposing superpower would be a signal for the possible use of nuclear weapons. So we would hope that such an invasion would be deterred. At the same time, it has clearly become accepted by both superpowers that certain aggressive actions would not involve the risk of nuclear retaliation. The Soviet Union has been able to get away with invasions of its Eastern European satellites and Afghanistan. The United States moved over half a million troops into Vietnam without risking retaliation. However, there remain situations where the probable response remains unclear. In 1962, Khruschev decided to move nuclear missiles to Cuba. He clearly did not foresee that this would be taken as a highly provocative act by the United States and one which was considered to directly threaten American national security. The United States responded with a blockade and the Russians stood down, but not before the world had come to the brink of nuclear war. One of the greatest threats to deterrence as a policy would be the emergence of another trouble spot (possibly the oil-rich Middle East) where the superpowers might misinterpret each other's vital national interests.

Perhaps the strongest argument against the deterrence theory, ("the cardinal weakness of the doctrine", in the words of Hedley Bull, a foremost student of the subject) is that it implies that both sides will act rationally at a time of crisis. Deterrence may be a good way of influencing long-term military planning, but it may prove completely inadequate in times of crisis. Throughout history, the decision to go to war, or to prepare for war, has often been made at a time when national emotions and suspicions are riding high. The fear of humiliation or the desire to act first, before a potential aggressor can hit you, may act to destroy the steady nerves which are essential in times of crisis. No one can predict how leaders will react, faced with the awesome responsibilities of power at such times.

The possibility that deterrence might fail, due either to miscalculation or to irrational behaviour by national leaders at a time of crisis, warns us how unwise it would be to continue to rely on deterrence as a long-term strategy. Its weaknesses as a policy are all too obvious and not much optimism for the future can be held just because deterrence has appeared to keep the peace between the superpowers in the past. As Lawrence Freedman has put it:

> An international order that rests upon a stability created by nuclear weapons will be the most terrible legacy with which each succeeding generation will endow the next. To believe that this can go on indefinitely without major disaster requires an optimism unjustified by any historical or political perspective.

It is thus a matter of urgency to search for alternative policies.

One of the greatest signs of hope in recent years has come from a renewed revulsion against nuclear war, a great popular outburst of protest against the continuing arms race and the threat of nuclear destruction. In Europe, it was the NATO decision to accept a new generation of tactical nuclear weapons which sparked off the protests. In Britain extra impetus was given by the decision to continue and even expand the British nuclear deterrent by replacing the existing Polaris missiles with Trident. In demonstrations all over Europe, hundreds of thousands have marched in opposition to nuclear war. The call from groups all over Europe was for disarmament.

Disarmament can come in three different ways. One is multilateral disarmament, when a number of nations come together and make mutual agreements to reduce their stocks of arms. Another is bilateral disarmament. This is when two nations agree to reduce arms. An example would be an agreement between the superpowers to reduce their weaponry. These forms of disarmament will be discussed in the next chapter.

Unilateral Disarmament

The third approach to disarmament is unilateral

disarmament. This is when one nation reduces its scale of armaments (or abolishes certain forms of armaments, such as nuclear weapons), irrespective of the response of its potential enemies.

There have been three major arguments put forward for unilateral nuclear disarmament. The first is that the possession of nuclear weapons, even only as a deterrent, is so morally wrong that a nation should not rely on them in any circumstances. This response forms an important element of the policy of the influential Campaign for Nuclear Disarmament (CND) and many Church leaders. The British Council of Churches, for instance, reaffirmed in 1979 a resolution passed in 1963,

> that nuclear weapons are an offence to God and a denial of His purpose for Man. Only the rapid progressive reduction of these weapons, submission to strict international control and their eventual abolition can remove this offence. No policy which does not explicitly and urgently seek to realise these aims can be acceptable to Christian conscience.

The immorality of nuclear weapons rests on their overwhelming destructive power, which threatens combatant and non-combatant alike, and on the likelihood that nuclear war would become uncontrollable. Such weapons can also be condemned on the grounds of their cost, although this is an argument which applies even more to conventional weapons. A reliance on nuclear defence can, in fact, reduce the total costs of a defence budget. This cannot mask the fundamental moral point that all arms expenditure can be seen as immoral in a world with so many other urgent needs.

The second argument for unilateral disarmament is that it will reduce the risk, or abolish it altogether, of nuclear war. As Betty England has written in the CND pamphlet "Nuclear Disarmament for Britain: Why we need action not words":

> The first reason why unilateral nuclear disarmament is the right policy for Britain is that the country would be a great deal safer without nuclear weapons than it is with them. We are told that these weapons are to deter a Soviet threat. But the threat to Britain would not exist if these weapons did not exist.

Betty England seems to be arguing that it is the presence of nuclear weapons in Britain which makes the country more vulnerable to nuclear attack. This would certainly be the case if the weapons could be destroyed with certainty by Soviet missiles, as, at a time of crisis, the Soviet Union would have every incentive to knock them out before they were used against it. However, as

Another face of the peace movement — a Japanese Buddhist chants prayers during a vigil outside the base at Greenham Common in Berkshire where Cruise missiles are to be kept.

we have seen, Britain has a stock of invulnerable missiles in its Polaris submarine missiles. The question remains: if Britain retains these nuclear weapons, which are clearly invulnerable to attack, is the country more or less likely to be the object of nuclear attack? It is difficult to see why it should be more so. The risks of retaliation would surely make the Soviet Union or any aggressor think twice.

It is difficult to see why, if Britain did abolish its nuclear weapons, this would make it necessarily safe from nuclear attack. We know, from the experience of Hiroshima, that it is possible for a nation at war to justify the dropping of nuclear weapons on an unprotected nation, though it would be wrong to rely too heavily on this argument in the very changed circumstances of the 1980s. At a time of war, Britain's important strategic position between the superpowers would make the country a possible target for destruction, whether it was defended by nuclear weapons or not.

A third argument for unilateral nuclear disarmament is that if one nation set an example, others would follow. As Martin Ryle has argued in his *Politics of Nuclear Disarmament*, British unilateral nuclear disarmament

> only makes full sense, and can only be properly advocated, if it is seen as integral to a wider and longer-term strategy which seeks to change the shape of NATO, to influence other states within and outside that alliance, to build a nuclear-free Europe and to strengthen popular disarmament movements everywhere.

Of course, it is very difficult to assess what the effects on its allies or opponents would be if Britain did renounce its own deterrent and the nuclear bases of the United States based on British soil. For the Soviet Union, the end of British nuclear weapons would make little difference to the total number of weapons aimed against it, as they form such a small part of this total. Similarly, if Britain's allies continue to see the Soviet Union as a threat, they might feel less rather than more secure as a result of Britain's renunciation of weapons and increase their defence programmes to compensate.

The arguments for unilateral nuclear disarmament remain attractive. As we shall see in the next chapter, progress in multilateral disarmament has been virtually non-existent. Unilateral disarmament presents a real chance to set the ball of disarmament rolling, a brave and positive attempt to end the risks of nuclear war. It remains, however, difficult to see how unilateral disarmament would necessarily make Britain a safer place to live in. In many ways, it would make it more vulnerable.

Despite the many flaws in the doctrine of nuclear deterrence, it does seem, on balance, to offer greater security for Britain than the alternatives. This deterrent need not, however, continue in its present form. There seems every argument for reducing the deterrent to a small number of invulnerable (probably submarine-based) missiles. It remains to be decided whether this deterrent should be under NATO control, or that of a non-aligned Europe, or possibly independently controlled under one country. The Europe option seems most attractive if the immensely difficult political problems involved in "decoupling" Europe from the United States could be solved.

If a reduction of nuclear weapons did take place, leading to the maintenance of a minimum deterrent force, it is probable that Europe would have to increase its conventional forces considerably in order to maintain an adequate level of security. This would certainly be expensive and it needs to be remembered that one result of nuclear disarmament might well be an increase rather than a decrease in European defence budgets. This would be the case unless one was certain that the long-term threats to the security of Europe offered by the Soviet Union or any other adversary were very slight. It would be a brave person who could make that prediction with any certainty.

In the long run, however, these are not solutions. We have outlined in this chapter the fundamental weaknesses of any policy involving nuclear weapons. Somehow, we have got to break the present moulds into which defence policy has settled. Stepping back from the national perspectives in which defence programmes have continued to be made, we can see the absurdity and danger inherent in the present situation. The great hope is that there is, at last, a significant groundswell of informed public opinion refusing to accept the risks of mass destruction and able to enforce international solutions to the problem.

There are no easy ways out of the present situation. There is no policy which does not offer risks. It is difficult, however, to see any way forward which does not include, first some general measures of multilateral disarmament, including nuclear weapons, and, secondly, a much more coherent and effective method of international peacekeeping. In the long run, it is only by eliminating fears and insecurities at national level that we can expect nations to reduce the arms behind which they at present shelter. How this might be done will be the subject of our final chapter.

A WAY AHEAD?

The Failure to Disarm

In 1978 the United Nations held a special session on disarmament. Looking back at the ten previous years they had this to say:

> The Disarmament Decade solemnly declared in 1969 by the United Nations is coming to an end. Unfortunately, the objectives established on that occasion by the General Assembly appear to be as far away today as they were then, or even further because the arms race is not diminishing but increasing and outstrips by far the efforts to curb it. While it is true that some limited agreements have been reached, "effective measures relating to the cessation of the nuclear arms race at an early date and to nuclear disarmament" continue to elude man's grasp. Yet the implementation of such measures is urgently required. There has not been either any real progress that might lead to the conclusion of a treaty on general and complete disarmament under effective international control. Furthermore, it has not been possible to free any amount, however modest, of the enormous resources, both material and human that are wasted on the unproductive and spiralling arms race.

There has been so much talk about disarmament in the twentieth century and yet so little appears to have been achieved. Why has this been so? Surely the dangers of the arms race are so obvious that mankind will be more than ready to make agreements to reverse the trend.

In his book, *Approaches to Disarmament*, Nicholas Sims outlines the reasons why he believes disarmament has been so difficult to achieve. First, there is the problem of agreeing as to which forms of weapons are to be involved in each set of negotiations. Should nuclear disarmament be the priority, or is the worldwide trade in conventional arms the real threat to mankind which needs to be dealt with first? If two sides to negotiations are both willing to reduce arms on a one-for-one basis, how do you measure one type of weapon against another, or ships against aircraft? From the 1920s onwards it has, in fact, been issues such as these which have prevented many disarmament talks getting under way at all.

A second reason for difficulty is, not surprisingly, the continued suspicions between nations. If an agreement is made to reduce arms, then each side must trust the other to carry out its side of the agreement. All too often this trust has not existed and so no agreement gets made at all.

There are other barriers to disarmament. There is what Sims calls "familiarity of armaments, unfamiliarity of voluntary disarmament". We are so used to carrying the risks of armaments that we can all too easily believe that the risks of disarmament, a process which is totally unfamiliar to us, must be greater. As Sims puts it, "People naturally tend to prefer the risks they know to risks they cannot so well imagine". Sims also mentions lack of sufficient political and public demand. When he was writing in the 1970s, this was certainly true. It is extraordinary to find, for instance, that between 1965 and 1980 there was not one debate in the British House of Commons on Britain's nuclear deterrent.

However, as we have seen, recent decisions in Europe, to place a new generation of tactical nuclear weapons on European soil, and in Britain, to replace its own Polaris nuclear missiles with the much more powerful Trident missiles, have led to an enormous upsurge of protest. The Campaign for Nuclear Disarmament, which had almost totally vanished from view after some years of popular support in the 1950s, has now re-emerged to play an important role in expressing public disquiet. It has cooperated with END (the European Nuclear Disarmament Campaign), which calls for a nuclear-free Europe "from Poland to Portugal", and with the World Disarmament Campaign, which lays more stress on multilateral disarmament (as compared to the unilateral disarmament policy of CND).

This revival of public feeling coincides with a new Special Session of the United Nations on Disarmament, opened in 1982. To fuel debate for this session, a group of influential world politicians and diplomats came together, under the Chairmanship of Olof Palme, Prime Minister of Sweden, to work out a practical plan for disarmament. Their report, *Common Security, A Programme for Disarmament*, is an impressive statement on the threats offered by the arms race

The United Nations opens its Second Special Session on Disarmament in 1982. Despite the worldwide interest in disarmament, these sessions have achieved little.

and it concludes with positive recommendations for change.

The Report gives a vivid picture of the threat of war, both nuclear and conventional. It details the highly destructive nature of nuclear weapons, the persistent tension between East and West, and the many conflicts which have broken out in the Third World, both within states and between states, since 1945. It describes the consequences of nuclear war, if one were to break out, without minimizing the very extensive damage which conventional or non-nuclear weapons can also cause.

Three main areas are pinpointed as of special concern. First, the position of the superpowers is examined. As they possess 95 per cent of the world's nuclear weapons and are the leaders of the two biggest military alliances in the world, any serious programme of disarmament must begin with them. Secondly, the Report concentrates on the situation in Europe. "Here", it says, "we find the greatest concentration of conventional and nuclear military power anywhere in the world. If war were to break out, it would be

UN forces have been used to help keep the peace on a number of occasions since 1945. Here a UN contingent, mainly of Pakistani soldiers, lands in Western New Guinea in 1962.

in this great centre of civilization that the greatest and most devastating destruction would occur". Third, the Report looks at the tensions of the Third World. Not only have these nations been building up their armaments at an ever more rapid pace, but it is in the Third World that most conflicts take place. The United Nations, which has the responsibility through the Security Council of maintaining world peace, has achieved very little. Certainly, on several occasions — during civil unrest in Cyprus, a breakdown of law and order in the Congo (now Zaire), and during the Suez Crisis of 1956 — the United Nations has sent in forces to try to prevent further fighting or to maintain a ceasefire. These forces, however, have had very little power. In the Lebanon in 1982, a United Nations peacekeeping force simply had to stand aside as an Israeli army

UN troops swept aside. Two UN soldiers in 1982, unable to do anything to prevent the Israeli invasion of the Lebanon, end up as prisoners in the custody of Lebanese Christian militiamen. Here they wait to be exchanged.

invaded the country. Across the Atlantic in New York, the United Nations could only pass resolutions calling on the Israelis to withdraw, resolutions which the Israelis ignored.

Having pinpointed these three particular areas of concern, the Report develops an important and comprehensive set of recommendations. It takes as its ultimate aim "general and complete disarmament". First, it tackles nuclear weapons and argues that not only must the strategic (or intercontinental) nuclear weapons be reduced by agreement between the superpowers, but the threshold — the point at which tactical, or short-range weapons, might be used — must be raised considerably.

> The world must break a system which equates the maintenance of peace with holding millions of human beings and the fruits of their labours as hostages for the good behaviour of the nuclear weapon states.

We have already explained (page 54) why many see tactical nuclear weapons as a specially worrying threat to peace. One valuable suggestion put forward in the Report is that nuclear weapons should not be stationed in the zone 150 kilometres either side of the German border. This would not, of course, prevent longer-range nuclear weapons being used if war did break out, but it might help delay the moment when their use is considered.

Alongside gradual nuclear disarmament by mutual agreement between the superpowers, the Report also recommends that there should be no further improvements in the quality of nuclear weapons. One of the most unstable elements of the arms race has been the introduction of new technology, offering perhaps a small increase in accuracy or destructive power of a particular weapon. The other side responds with similar "improvements", and so the "race" goes on, with each side trying to steal an advantage over the other. As part of the attempt to restrict technological improvement, there also needs to be a comprehensive ban on all nuclear testing. Most forms of tests were banned by agreement in 1963, but SIPRI calculates that 264 tests took place between 1975 and 1980.

The Report recognizes that few moves will be made towards full disarmament until all nations, and particularly the smaller nations of the Third World, feel more secure. Common security can only be achieved by international agreement and cooperation and the Report goes on to concentrate on ways in which the United Nations might be strengthened in its role.

The present weakness of the United Nations stems from the failure of the members of the Security Council to be able to pass resolutions which they would then be able to implement, by force if necessary, if they were not complied with.

As we have mentioned, any one of the five permanent members of the Security Council (the United States, the Soviet Union, France, Britain and China) can use its vote to prevent a resolution being passed. All too often the tensions of the Cold War have spread to the Council, members of one "side" voting against any resolution which they feel might offer some advantage to the other.

The Report recognizes that these suspicions will not disappear easily and it is not enough just to call on the permanent members of the Security Council to act more responsibly. However, it does hope that the members might begin to cooperate more effectively on issues involving Third World states, where, as we have seen, conflict is most prevalent. The Report also calls for more effective action by the Secretary-General of the United Nations. He should be required to present regular reports to the Security Council on the state of tension in the world.

The Report thus enshrines two concepts: first, moves towards general disarmament, and secondly, an improvement of common security as the result of a more effective role exercised by the United Nations. If needs be, the United Nations will have to use force to keep the peace and another recommendation of the Report is that nations should offer troops on a stand-by basis to help react to threats to the peace. Such forces are, in fact, already provided for in the Charter of the United Nations, but they have never been offered on a permanent basis by member states.

There are many other recommendations in this important report and it deserves to be read in full. It remains to be seen whether it will be one of the many similar reports which simply end up gathering dust, or whether it will be a significant force for real change.

Some hope that it might be the latter comes from the increasing strength of public opinion. In the past few years, it has become clear that important sections of public opinion in Western Europe and the United States have lost faith in conventional defence policy. There is far greater critical examination of any move which involves the increase of armaments or so-called improvements in their quality.

If this pressure of opinion is sustained, it is inevitable that politicians will have to take it into account. There are already signs that Western leaders are approaching decisions about defence policy more cautiously. President Reagan of the United States, for instance, has begun talks with representatives of the Soviet Union on limiting tactical nuclear weapons in Europe (the so-called zero-option by which the Soviet Union would

Olof Palme, Prime Minister of Sweden, whose report, *Common Security*, marks an important attempt to start a process of disarmament and common security for mankind.

remove their SS-20 missiles and the West stop deploying their new tactical weapons). There are some doubts as to the commitment of the United States to these talks, especially in the light of increases made in the recent defence budgets, but they are a welcome sign that political pressure may have results.

The only real way to end the threat of nuclear war is by effective international control of nuclear weapons. At present, there is little sign of a world body emerging with sufficient power to achieve this control. We are left with the dilemma of how to design a defence policy to cope with opponents who may retain their nuclear weapons. It is an area in which we are still inexperienced. Our knowledge of how to run a defence policy using the deterrent effect of nuclear weaponry is very limited and is based only on one example, the confrontation between the forces of East and West in Europe. Whether nuclear deterrence could even be considered as a valid policy in any other situation is doubtful.

The only realistic response to this situation is to build up public awareness across nations of the dangers confronting mankind. It will only be when the world community comes to realize that the sum of individual defence policies designed to achieve security for one nation is no more than the increased insecurity of mankind in general, that effective forces for change may emerge.

The Commission on Disarmament and Security sums it up well.

> The destructive power of modern nuclear and conventional weapons, both in quality and quantity, has totally outrun traditional concepts of war and defence. In the event of a major world war, which would escalate inexorably to the use of nuclear weapons, all nations would suffer devastation to a degree that would make 'victory' a meaningless word.

The threat of nuclear war and the continued suffering in conventional war of large areas of the world make it all too clear how urgent it is for us to break the moulds of traditional defence policies. This will never be easy. Deep-rooted suspicions, strong national prides, the force of inertia will all combine to make the necessary substantial changes difficult to achieve. At last, there are signs that world leaders are aware that the collective dangers of their national policies outweigh the temporary and shaky security which each policy is designed to bring. It is the hope of this book that public pressures and the enlightened leadership of major world politicians will together bring about a change towards a more responsible global policy, involving the reduction and eventual abolition of nuclear weaponry and an effective development of common security.

GLOSSARY

aggression an action involving armed attack by one nation on another.

appeasement a policy followed by Britain and France in the 1930s, by which they hoped to prevent war with Germany by coming to agreements with Hitler over points of conflict. Hitler had no intention of keeping his agreements and the policy failed to keep the peace.

arms control agreements to limit the number of arms held by one or more nations. A fixed number of a certain type of weapon might be set as the upper limit allowed. Arms control does not necessarily mean there will be a reduction in armaments.

Brandt Report an important report on International Development issued in 1980 and known after its Chairman, Willy Brandt. It calls for a positive programme of action by the richer nations of the world to help the poorer.

CND the Campaign for Nuclear Disarmament, founded in 1957. CND believes that Britain should unilaterally (see *disarmament*) disband its nuclear weapons. After some years of decline, CND has re-emerged in the 1980s as a major political force.

Cold War the state of tension which has existed since the Second World War between the United States with its allies (this includes Western Europe) and the Soviet Union with its allies. Each side has built up enormous stocks of nuclear weapons, the existence of which, it hopes, will deter the other from attacking.

collective security a policy by which a number of nations agree that they will cooperate to prevent war. They might, for instance, agree to provide troops to punish an aggressor. The United Nations is an important collective security organization.

communism a political system which believes in the formation of a community in which all will live and work together on an equal basis. Many so-called communist nations in the present world accept that this equality has not as yet appeared, but they claim that they are still working towards it.

conventional weapons a term used to describe all non-nuclear weapons, such as tanks, ships and aircraft.

Cruise missile see page 44.

deployment A weapon is *deployed* when it is put into position ready for use in attack.

deterrence the name of a policy which aims to prevent aggression by maintaining large stocks of weapons (usually nuclear), which would be used to destroy any attacker. (See pages 52-57).

diplomacy the art of peaceful negotiation of differences between nations.

disarmament any policy which results in the reduction of the numbers or power of armaments. Multilateral disarmament occurs when a number of nations agree to reduce their arms. Bilateral disarmament occurs when two nations agree. Unilateral disarmament is when one nation reduces its armaments without making any agreement with other nations that they will do so.

END the European Nuclear Disarmament campaign, which campaigns for a Europe free from nuclear weapons.

fascism a political system which believes in the glorification of a nation or race, normally under the strong leadership of one individual. Fascist states normally glorify war, seeing it as a means of proving the superiority of their nation or race. Hitler's Germany and Mussolini's Italy are the main examples of fascist states.

flexible response a military plan by which an attack is met by a variety of military responses, ranging from the use of conventional weapons to nuclear weapons, according to the seriousness of the attack. It is the policy at present followed by NATO in Europe.

Geneva Conventions a series of agreements signed in 1949 but with later additions, which lay down rules for the humane conduct of war. They have been devised by the International Red Cross.

guerrilla warfare war conducted by small bands of armed men and women against a conventional army — which can be that of their government or of an invader. Guerrilla wars normally take place in rural areas where the guerrillas can hide and carry out surprise attacks and ambushes. It is extremely difficult to destroy well-organized guerrilla forces.

independent nuclear deterrent nuclear weapons held under the control of one nation and liable to be used, in the last resort, on the decision of the leaders of that nation. The two nations with an independent nuclear deterrent in Europe are Britain and France.

insurgency a state of continual unrest, usually involving a rebellion or uprising, which continues over a period of time.

just war war which is believed to be morally just. This may be because it is war waged to protect a helpless victim, to punish an aggressor or to achieve fundamental political rights. There have been considerable problems in defining the circumstances in which war is "just".

League of Nations an international organization set up at the end of the First World War in the hope of preventing further outbreaks of war. It lacked sufficient will among its leading members to act as an effective force for peace and was dissolved at the end of the Second World War to be replaced by the United Nations.

massive retaliation a military plan evolved by the United States in the 1950s by which an attack on Europe by the Soviet Union was to be met with a massive nuclear counter-attack. The plan was dropped when the Soviet Union developed its own nuclear weapons in large quantities and could threaten the United States in return.

NATO the North Atlantic Treaty Organization, founded in 1949, a military alliance which includes most nations of Western Europe as well as the United States and Canada. It aims to provide them with a common means of defence in case of attack by the Soviet Union or its allies.

Non-proliferation Treaty a treaty which came into effect in 1970 and which hopes to limit the spread of nuclear weapons. The nations which possess nuclear weapons promise to reduce their own stocks of nuclear weapons and the nations which do not have nuclear weapons promise not to obtain them. There have been numerous loopholes in this treaty. Two nuclear nations, China and France, have not signed it.

nuclear weapons There is a wide range of nuclear weapons. These are described in some detail in John Cox's book *Overkill* (Pelican 1981). For Cruise missiles, Polaris and Trident see pages 44, 48-51. An important distinction is between strategic nuclear weapons and theatre, tactical or battlefield weapons. Strategic nuclear weapons are intercontinental — for instance, missiles in the United States aimed at the Soviet Union and vice versa. They tend to be aimed at cities and have great destructive power. Tactical, theatre or battlefield weapons are those based in Europe aimed at the Soviet Union, or those in the Soviet Union aimed at Europe. They are thus of shorter range and usually, though not always, of less destructive power. For more details of these weapons see pages 54-56.

pacifism the belief that the use of violence in any circumstances — even self-defence — is wrong.

peacekeeping forces forces which are introduced into situations where there is tension or conflict in the hope of maintaining the peace. The United Nations has provided such forces from among its members on several occasions in the past thirty-five years. There is often little such forces can do if conflict breaks out again, and often they have to be withdrawn if the nation in which they are based requests it.

Polaris a nuclear missile system based in a submarine. See page 48 for more details.

proliferation In defence policy, *proliferation* refers to the spread of weapons. The term is often used, in particular, to refer to the spread of nuclear weapons.

resolution a statement, often outlining a desired course of action, passed by an international or other organization. The United Nations might, for instance, pass a resolution calling on two warring nations to stop fighting.

SALT Strategic Arms Limitation Talks. These talks, carried out between the superpowers in the 1970s and 1980s, have been concerned with reducing their stocks of strategic nuclear weapons. Two major agreements have been made and there are now upper limits on certain types of nuclear weapons and defence systems.

sanctions a form of punishment, agreed by a number of states, against a nation which has offended international law. Sanctions may be diplomatic (cutting off all political relationships with the nation), economic (refusing to trade with it) or military (using armed force to attack it). Both the League of Nations and the United Nations Charters have provision for sanctions against states which conflict with the principles of these charters.

SIPRI the Stockholm International Peace Research Institute, an important organization which gathers information on all aspects of armaments and defence policy. A number of its findings have been included in this book.

superpower a word normally used to describe the United States and the Soviet Union, which, in resources, military strength and international influence, dominate the other, lesser "powers".

task force a military expedition gathered together to undertake a particular military task. The naval task force sent out to regain the Falkland Islands in 1982 is an example. This force (which included airforce units and army forces) was put together with one specific task in mind, a task which it achieved successfully.

Test-Ban Treaty a Treaty signed in 1963 by which the United States, the Soviet Union and Britain agreed only to test nuclear weapons underground. Two "nuclear" nations, China and France, have not signed it.

Trident a submarine-launched nuclear missile system. See page 49.

United Nations an international organization, founded in 1945, whose primary purposes are to improve international cooperation and maintain world peace. The main organ of the United Nations is the Security Council, which consists of fifteen nations, five of whom, Britain, France, China, the United States and the Soviet Union, are permanent members. The failure of these five nations to agree in the Security Council has made the United Nations a much less effective force for world peace than it might be.

war of liberation a war waged by insurgents within a territory to free themselves from a government or invading force which, they believe, is oppressing them.

Warsaw Pact a long-term military alliance of the Soviet Union with its Eastern European allies, signed in 1955. It confronts the similar NATO alliance formed by the Western nations in 1949.

RESOURCES LIST

There are a vast number of publications on nuclear weapons, defence and disarmament. Governments issue their own reports on Defence Policy; there are organizations which gather information for those concerned with these issues; and, of course, there are many books, pamphlets, etc on every aspect of the subject. This list includes those which this author has found useful or important.

BOOKS

General
Shelford Bidwell, *World War Three*, Hamlyn Paperbacks. A series of essays on developments in defence and how a Third World War may break out. Readable.
Nigel Calder, *Nuclear Nightmares*, Penguin Books. Gives details of the effects of nuclear attacks but believes there is a case for nuclear deterrence.
Field-Marshall Lord Carver, *A Policy for Peace*, Faber and Faber. One of Britain's more intellectual soldiers adds his own thoughts on the problems raised by nuclear weapons.
John Cox, *Overkill, The Story of Modern Weapons*, Penguin. Includes a chapter on the history of CND.
Lawrence Freedman, *Britain and Nuclear Weapons*, Macmillan Paperbacks. A good history of why Britain acquired and has continued to deploy nuclear weapons.
Lawrence Freedman, *The Evolution of Nuclear Strategy*, Macmillan. A large and expensive book, but the fullest and most balanced history of how nuclear strategy has evolved since 1945.
Laurence Martin, *The Two Edged Sword, Armed Force in the Modern World*, BBC. The Reith Lectures for 1981. Important reading for those concerned with the problem of strategy. Accepts, with reservations, the status quo, but deserves to be read and analysed by those who don't.
Robert Neild, *How to Make Up Your Mind About the Bomb*, Andre Deutsch. Paperback. Presents arguments for and against the present reliance on nuclear weapons, finding more against than for.
New Statesman Papers No. 3., *Britain and the Bomb*, Paperback. A series of articles of very varying quality.
Paul Rogers and others, *As Lambs to the Slaughter, The Facts about Nuclear War*. Lecturers from the Bradford School of Peace Studies explore the dangers of nuclear war.
Anthony Sampson, *The Arms Bazaar*, Coronet Books. Slightly dated account of the international arms trade.
E.P. Thompson and Dan Smith, *Protest and Survive*, Penguin Special. A now famous collection of essays by those involved in the disarmament movement.
Solly Zuckerman, *Nuclear Illusion and Reality*, Collins. A former adviser to the British Government on Defence offers an excellent criticism of the arms race and presents an argument for a minimum deterrent.

Ethical Problems of War
The Church and the Bomb, Hodder and Stoughton. Report on the Anglican Church's attitude to the bomb. Takes a unilateralist line.
Jonathan Glover, *Causing Death and Saving Lives*, Penguin. A philosopher looks at the arguments involved in all cases in which life is taken or threatened, including war.
Michael Howard, *War and the Liberal Conscience*, Oxford University Press. An excellent examination of how "liberals" have faced up to the dilemmas involved in the use of war.
Michael Walzer, *Just and Unjust Wars*, Penguin. An excellent overview, with a wide variety of historical examples, of the ethical problems involved in the fighting of wars.

Disarmament
Mary Kaldor and Dan Smith, *Disarming Europe*, Merlin Press. A series of essays on the problems involved in European disarmament.
Olof Palme (Chairman), *Common Security, A Programme for Disarmament*, Pan Books. An independent report by a high-level Commission on Disarmament and Security issues. Essential reading.
Martin Ryle, *The Politics of Nuclear Disarmament*, Pluto Press. Argues for disarmament through the pressures of popular protest.
Nicholas Sims, *Approaches to Disarmament*, Quaker Peace and Service. Probably the best introduction to the problems involved in achieving disarmament.

INFORMATION

Three organizations, among several, deserve mention:

International Institute for Strategic Studies, 23 Tavistock St, London WC2E 7NQ. Broadly supports the status quo. Publishes a bi-monthly magazine, *Survival*, as well as a series of pamphlets, the Adelphi Papers, which cover a wide variety of defence issues. Also yearly reports, *The Strategic Survey* and *The Military Balance*.
Stockholm International Peace Research Institute issues a Yearbook, *World Armaments and Disarmament*. A reliable and respected source.

The Armament and Disarmament Information Unit (based at the University of Sussex, Falmer, Brighton BN1 9RF) provides factual information for those who need it.

FILMS

Concord Films, 201 Felixstowe Road, Ipswich, Suffolk IP3 9BJ have a good selection of films on war and peace. They now have a separate catalogue of peace films.

ORGANIZATIONS

CND, the Campaign for Nuclear Disarmament, 11, Goodwin St, London N4 3HQ.

END, European Nuclear Disarmament, 227, Seven Sisters Road, London N4.

Pax Christi, the Catholic organization for peace, St Francis of Assisi Centre, Pottery Lane, London W11 4NQ.

The Peace Education Network is particularly concerned with the spread of peace education in schools. Membership Secretary c/o Centre for Peace Studies, St. Martin's College, Lancaster LA1 3JD.

Quaker Peace and Service, Friends House, Euston Road, London NW1. The Quakers have made an impressive contribution to peace movements and they have produced a number of publications, etc.

The School of Peace Studies, Bradford University is an important centre for the study of means of resolving conflict.

INDEX

Afghanistan, Soviet invasion of (1979) 37
appeasement 24-25
arms sales 19
atomic bomb *see* nuclear weapons

Berlin
    emergence of East and West Berlin 33
    crisis of 1948-49 34
    crisis of 1959-61 36-37
Brandt Report, quoted 19
British Council of Churches, resolution against nuclear weapons 58
British defence policy *see* United Kingdom

Campaign for Nuclear Disarmament (CND) 51, 58, 61
Carter, Jimmy, President of United States 1976-80 55
Chamberlain, Neville, British Prime Minister 1937-40 24
Chevaline (warhead) 49
China
    and links with the West 40, 41
    Mao Zedong achieves power 35
    and nuclear weapons 17
    permanent member of United Nations Security Council 7
civilians in warfare 14-16
Cold War 26, 32-39
Common Security, A Programme for Disarmament (Palme Commission Report) 1982, discussed 62-65
    quoted 9, 16, 66
conventional weapons 47
    British defence policy and 45
    cost of 17-19
Cruise missiles 44-45
Cuba
    Bay of Pigs invasion (1961) 37
    missile crisis (1962) 37, 57

deterrence, theory of 20, 52-57
disarmament 57-60, 60-65

Eden, Anthony, British Prime Minister 1955-57 25-26
European Nuclear Disarmament Campaign (END) 61

Falkland Islands crisis (1982) 13, 19, 24, 28, 29, 46
fascism 4, 12
    British policy towards 11, 24-25
    and the "just" war 11
    rise of 11
First World War 3, 11
France
    and nuclear weapons 17, 47
    permanent member of Security Council of United Nations 7

Geneva Conventions (1949, 1977) 16

Helsinki Agreement (1975) 40
Hiroshima 5, 7, 16
Hitler, Adolf, German fascist leader 31-32 (*see also* fascism, Second World War)

Independent Commission on Disarmament and Security Issues (Palme Report) 1982 *see* Common Security
India
    defence expenditure 19
    and nuclear weapons 17

Japan
    policy in 1930s 29
    Second World War 5
    (*see also* Hiroshima)
"just" war 11-13, 19

Kellogg-Briand Pact (1928) 4
Kennedy, John, President of the United States, and Cuban missile crisis 37
Korean War (1950-54) 36
Kruschev, Premier of the Soviet Union 36-37

League of Nations 4, 6, 11

Middle East 9-10, 27
    crisis of 1973 22
    Suez Crisis (1956) 25-26

National Security Council Report 68 (NSC 68) 35
NATO (North Atlantic Treaty Organization) 21-22, 36
    Britain and 44, 46-47
    United States and 22-23
Non-proliferation Treaty (1970) 17, 50

nuclear strategy 20, 52-60 (*see also* deterrence)
nuclear weapons
    atomic bomb dropped on Hiroshima 5
    British nuclear deterrent 47-51
    effect of 5, 16-17
    numbers of 17
    strategic nuclear weapons 17, 54
    tactical nuclear weapons 52-56, 64
    (*see also* Cruise, Polaris, Trident, SS-20)

Palme, Olof, Chairman of Commission on Disarmament and Security 62, 65
Poland, after Second World War 32
Polaris missile 48-49
Presidential Directive 59 (1980) 55

Reagan, Ronald, President of the United States 23, 26
Red Cross 16
Russia *see* Soviet Union

SALT (Strategic Arms Limitation Talks) 54
Second World War 5
    bombing raids in 16
Soviet Union 30-41
    Afghanistan invasion 37
    Berlin crises 33, 34, 36-37
    Cold War 32-34
    Cuban missile crisis 37
    Ethiopia 38-39
    and Hitler 31-32
    under Kruschev 36-37
    military aid to Third World 40
    military strength 39-40
    number of weapons 17
    under Stalin 31-36
    use of military force by 10
Spanish Civil War 26
SS-20 (missile) 44
Stalin, Josef, leader of Soviet Union 31-36
Suez Crisis (1956) 24-25

tactical nuclear weapons 54-56, 64
    (*see also* Cruise)
Trident (submarine-based missile) 49-51
Truman, President of United States 35
    quoted 26

unilateral disarmament 57-59
United Kingdom
    bombing raids in Second World War 16

71

expenditure on defence 42
and Falkland Islands crisis 13, 19, 24, 28, 29, 46
and fascism 11, 24-25
in NATO 44, 46-47
naval forces 44, 45, 46
nuclear bases 44
nuclear deterrent 47-51
and Polaris 48-49
and Suez Crisis 25-26
and Trident 49-51

United Nations 6
   and collective security 23, 64-65
   and disarmament sessions 61-62
   and "just" war 13
   peacekeeping forces 63-64
   powers of 23
   Security Council 6-7, 23
United States
   attitudes to communism 22, 26-27, 36
   bases in Britain 44
   and coming of Cold War 26-27, 34-36
   Japanese attack on (1941) 5
   and NATO 22-23
   Presidential Directive 59 55
   and use of military force

Warsaw Pact 21, 36
Washington Treaties (1923) 14
weapons *see* conventional weapons, nuclear weapons